Dale McIntosh

The Long Distance Feeling

Elaine Bernard

The Long Distance Feeling

A History of the Telecommunications Workers Union

NEW STAR BOOKS
Vancouver

First printing November 1982
5 4 3 2 1

Canadian Cataloguing in Publication Data

Bernard, Elaine, 1951-
The long distance feeling
Includes index.
ISBN 0-919573-02-9 (bound). — ISBN 0-919573-03-7 (pbk.)
1. Telecommunications Workers Union (British Columbia) — History. 2. Trade-unions — Telephone workers — British Columbia — History.
3. Telephone — British Columbia — History.
4. British Columbia Telephone Company — History.
I. Title.
HD6528.T32T44 331.88'113846065711 C82-091334-0

Typeset and printed by union labour
Manufactured in Canada

The publisher is grateful for assistance provided by the Canada Council

New Star Books Ltd.
2504 York Avenue
Vancouver, B.C.
V6K 1E3

Contents

Acknowledgments

In writing this book I accumulated many debts, the greatest of which I now have an opportunity to acknowledge. My thanks to Lindsay Allen, whose work on the early history of BC Telephone was of great assistance; Peter Anderson and Jean McNulty of the Telecommunications Research Group at Simon Fraser University; Rick Dowling and Cliff Rundgren of the International Brotherhood of Electrical Workers Local 213; John Smart of the Public Archives of Canada; Lanny Beckman and Ralph Maurer of New Star Books; my editor, Julian Beder; and to Bill Clark, president of the TWU, Bob Donnelly, past president of the TWU and the originator of this project, Greg Gigg, Bert and Claudine Johns, Bob Rogow, Chris Schenk and Sid Shnaid. A special word of thanks to the staff at the Telecommunications Workers Union office, members of the Office and Technical Employees' Union Local 15; to the TWU business agents; to the staff at the Vancouver City Archives; and to the staff of the University of B.C. library's special collections branch (now home to the TWU archives), especially George Brandak.

I would like to dedicate this book to the members of the Telecommunications Workers Union, in recognition of the battles they have fought and in the hope that it will aid them in the struggles to come.

Introduction

When members of the Telecommunications Workers Union occupied telephone exchanges across the province in February, 1981, they made national news. For most workers the tasks they performed during the occupation were much the same as their daily routine. But there was a significant difference. The workers were collectively controlling and operating the telephone network without management. For the media, a cross-province occupation by striking telephone workers constituted one of the most dramatic labour confrontations in recent history.

The 1981 occupation was the culmination of a conflict which went back many years. The present cycle of labour unrest began when BC Telephone workers struck in 1969. In the wake of that strike came further confrontations between union and company, including a strike at Okanagan Telephone and a long province-wide lockout. Industrial relations at BC Telephone were in turmoil. An accurate parallel is the comparison of BC Telephone labour relations to those prevalent at the post office, the scene of a decade-long battle over technological change and arbitrary moves by management. As with the post office employees, telephone workers had witnessed extensive technological change introduced by management without consultation with the workers whose jobs had been profoundly affected.

The TWU is well aware of the significance of technological change. Technological change has always affected its members and has spurred actions by the union, including the 1981 occupation. But the early history of work in the industry and the continuous impact of technology on the telephone workers have never been systematically researched. This history exists only in the memories of retired members.

Months before the occupation, a convention of the TWU decided to commission a history of telephone workers in British Columbia. They wanted the history to be as complete an account as possible, a "warts and all" history which would help them understand their past and educate new members as well. This book is the product of their decision. It is the story of the telephone workers, their work, and their union.

It might seem odd that a union would have to go to the length of commissioning a history in order to discover the history of working people in the communications industry. But history as it is taught in schools is the story of politicians, captains of industry, and the occasional political crisis. This type of history holds little relevance for the majority of people. It is not surprising to find that most books written on the communications industry in Canada are about entrepreneurs and owners of companies, government regulation of radio, television and telecommunications, or glossy renditions of how Bell or GTE is bringing telecommunications to the Canadian wilderness. There is little mention of the people who built, operated and maintained the telephone system. These people, their lives, their organizations, their successes and their failures are an integral part of the history of this country, but a part which has been ignored for too long.

The lack of studies on workers in the communications and electrical industry is especially critical today because society is at the beginning of a new industrial revolution. It is important for workers today to gain the perspective of their predecessors in the electrical industry, not because conditions of yesterday will apply again in the future, but because an understanding of where we came from and how and why things change is essential in equipping ourselves for the future.

British Columbian telephone workers have been organized longer than most North American telephone workers. Much of their past parallels the history of British Columbian and Western Canadian labour movement. Telephone workers first organized at the turn of the century as a militant local of an international union. After initial successes the telephone workers broke from the electricians' local and formed their own local which, under company influence, declined and eventually disappeared, to be replaced by a company union in the late twenties. With the change in labour legislation during World War II, the company union of the thirties emerged as an independent Canadian union. The late sixties saw a resurgence of labour militancy, with telephone workers going on strike for the first time in 50 years.

The telephone company itself has changed profoundly through its 100-year history. At one time nearly 50 small telephone companies existed in British Columbia. But by the end of World War I, BC Telephone had established its dominance and controlled 97 per cent of the telephones in the province.

Today, BC Telephone is the only Western Canadian telephone company which is not publicly owned. Like many industrial giants in Canada, it is foreign owned. Many older telephone workers trace the beginning of their dissatisfaction with BC Telephone to the sixties, soon after the company was taken over by the American multinational, General Telephone and Electronics (GTE). Yet BC Telephone has been foreign owned since 1926, when a United States utilities corporation, Theodore Gary and Company, bought the company. The significance of the GTE takeover, however, was that it marked changes in corporate management and strategy. Corporate changes led to changes in the attitudes of workers.

Technological change—a continuing theme in the telecommunications industry—has emerged as one of the most pressing issues for the whole labour movement. New microelectronics technology is revolutionizing the way we work, the way we are educated, and the way we spend our leisure time. One of the reasons that workers originally formed unions was to gain a collective voice in the workplace. But it is extremely difficult to negotiate the use of technology appropriate to the needs of workers. The choice of appropriate technology, and the organization of work, are claimed as prerogatives of management. Yet the selection and design of technology has a profound effect on all of us.

There is a growing contradiction in the communications industry, a contradiction inherent in the corporate use of new technology. The promise of the industry has been the power to shrink distances, to bring voices from afar, and to increase human contact through innovation. But for workers in the communications industry there has not been a closing of "that long distance feeling." Rather, that distance has increased. Today workers in the industry are more alienated from their work, from their customers, from management, and from each other than at any time in their history.

Telephone workers have developed a sense of professionalism and craft pride. While a number of craft jobs are peculiar to the telephone industry, workers in non-craft jobs—clerks, operators, and sales staff—have always seen themselves as part of a distinct profession—telephone workers. A number of workers are second- and third-generation telephone workers. Like their parents, they have sought jobs with the telephone company because it was seen as good, steady, important work. It was more than a job; it was a profession which provided a socially valuable service.

Much of the workers' traditional professional pride has disappeared over the last decade. In its place is a growing sense of alienation permeating every aspect of their work. There is now one management person for every three or four workers. In addition to the monitoring performed by this massive supervisory work force, electronic monitoring is now conducted on an unbelievable scale. To make matters worse, the managerial staff also serves as a built-in reserve pool of strikebreakers designed to deprive the union of its chief weapon.

Telephone workers' sense of unease is well-founded. The new electronic switching equipment has relegated a craft worker to a "card puller." Faulty equipment is rarely repaired; instead, components are junked. There are fewer and fewer operators; in their place are mechanical recordings. These will eventually be replaced by computer-simulated responses.

Understanding the new technology, who controls it, why it is introduced, and how it is implemented is vital to achieve an understanding of the past. This book offers some insight into the success and failure of different forms of organization, as well as the tactics and strategies that workers have used in the past. While these solutions cannot be mechanically applied to future situations, they offer an appreciation of the complicated processes and constant changes through which workers' organizations arose.

For telephone workers, technological change is not new. The telephone industry was born of technological change. The telephone was a technological innovation which replaced the telegraph as a cheap, rapid means of communications. For management, the telephone had the further advantage in that it required less skill of the operator than the telegraph. The industry is full of examples of new technology which brought about dramatic restructuring of work. Over the next few decades conflict and confrontation in the work place will focus to an even greater extent on this issue. The conflict will centre on workers' struggles to control the organization, design, and product of their own work. Microelectronics and related technology today provide the potential to make work safer and more interesting, increase production, and decrease work time. This potential will only be attained if new technology is introduced with the cooperative participation of workers and is used to aid workers in performing new jobs.

Chapter 1 *Early Years*

The telephone, like its predecessor the telegraph, was one of the first industries based on the commercial use of electricity. The commercial value of telegraphy had originally been demonstrated in the 1840s, when the United States Congress backed an experiment to build a telegraph line from Baltimore to Washington which successfully transmitted the message, "What hath God wrought!" Over the next twenty years telegraph lines were extended right across North America, reaching the Pacific Coast in 1861 and eventually England with the laying of the Atlantic cable in 1866.

The new industry gave birth to one of the most successful business enterprises of the period, the Western Union Telegraph Company. Founded in 1856, Western Union grew into a multimillion dollar corporation within a few years. The use of the telegraph also brought into existence a newspaper wire service, Associated Press, which made it possible to transmit news rapidly across the country.

The demands of the new industry required a new and specially skilled work force. Telegraph operators learned to code and decode messages rapidly into a system of dots and dashes, named Morse code after its originator, Samuel Morse, for transmission over the wires. Linemen, cable splicers and construction crews built and connected thousands of miles of line through cities, along railway tracks and through every imaginable terrain to span the continent with its first electrical communications network.

In 1876 an important new electrical invention made its debut at the Centennial Exposition of Philadelphia — the telephone. In rapid succession came a number of electrically powered instruments and transmission devices, as well as stations capable of transmitting electricity for commercial sale. These paved the way for electrically powered transportation, street railways, the incandescent lamp, and electric street lights. The Age of Electricity had arrived.

The pace of development of the telephone — from invention to commercial application — was remarkable. Alexander Graham Bell first patented the telephone in March, 1876, and in June of that year he demonstrated his new instrument at the Centennial Exposition. A year later he established the Bell Telephone Company, which launched the commercial development of the telephone.

The Bell Company of 1877 was a far cry from the gigantic corporation of today. Success in the telephone field was far from assured. From the start Western Union Telegraph, which controlled the telegraph industry in the United States, challenged Bell's patents. Western Union had the money, the experienced work force, and a national network of poles, wires and offices, which could all be marshalled to compete with the emerging telephone company. In one of the most important corporate battles in history, Western Union reached an out of court settlement with Bell. The telegraph company surrendered its telephone network to Bell in exchange for 20 per cent of Bell's telephone revenues during the seventeen-year life of the Bell patents. The withdrawal of Western Union opened the field for Bell's meteoric rise as the giant in the communications industry.[1]

In Canada Alexander Graham Bell turned over his interests to his father, Melville Bell. Following the lead of its sister company in the United States, the Canadian company leased telephone equipment instead of selling it. Bell senior set up a network of agents holding franchises for the Bell equipment. By January, 1879, there were 22 agents throughout Canada, including Robert Burns McMicking, Bell's agent for British Columbia.[2]

McMicking, originally from Ontario, had travelled west during the Gold Rush of 1862. He failed to make his fortune in the Gold Rush and returned to his trade, telegraphy. He worked for a few years in the interior of British Columbia on the Collins Overland Telegraph, a project which involved an attempt to build a telegraph link from British Columbia through Alaska, across the Bering Strait to Russia and Western Europe. The laying of the

Atlantic cable put an end to the scheme, and McMicking then settled in Victoria, where he became superintendent of the British Columbia (later Dominion Government) Telegraph System.[3]

While employed as superintendent of the telegraph system, McMicking heard of the telephone. He wrote to Melville Bell, requesting information and a pair of telephones. In the spring of 1878 McMicking received his telephones. After testing the sets by placing a line between his home and his office, he applied for and obtained a franchise which allowed him to lease telephones from Bell. McMicking began advertising his telephone service at the rate of $25 a year, but had few takers. For the next two years McMicking was forced to retain his full-time post with the telegraph company. As Victoria had no exchange, his telephones could operate only as intercoms.

In fact the first telephones on the Island were not even provided by McMicking and his Bell agency. In 1876 an article appeared in *Scientific American* explaining the principle of the telephone and how to construct one. William H. Wall, a mechanic at the Dunsmuir and Diggle Wellington Mine near Nanaimo, built a pair of telephones according to these specifications for use at the mine. They were used to communicate between the mine and the dock at Departure Bay a few miles away. This early telephone line was the first in regular use in the province.[4]

W. J. Jeffree constructed a second telephone line in Victoria in January, 1880, when he installed a pair of telephones connecting his clothing store to a soap factory owned by his cousin. The telephones, acquired on a trip to San Francisco, raised much interest. The local newspaper, the *Daily British Colonist*, even sent a reporter to cover the event, and the subsequent article suggested that this new means of communication represented the way of the future:

> The telephone line was completed from W. J. Jeffree's clothing house to Pendray and Co.'s soap factory yesterday, and many messages passed between the two points. The distance is about one-third of a mile. The mode of communication is very simple. In the office at each end of the wire there hangs against the wall a small black-walnut box. At the side of the box is a small crank, which a person desirous of communicating with another turns. The movement of the crank rings an alarm bell and the attention of the person sought being directed to the instrument, he either repeats the alarm by turning a crank at his end, or taking the speaking-tube in his hand, he places it near his mouth and asks what is wanted. The sound of the voice strikes against a diaphragm within the tube and is conveyed by the wire to the other end, and so on. A conversation thus commenced may be carried on as long as desired. Last evening a

> good many messages were dispatched and received by aid of the telephone from and at Mr. Jeffree's, the voices of those speaking at the other end being plainly recognized. A gentleman with a bad cold ventured into the office, and as he stood examining the machine he coughed. Instantly from the other end came the information, "That's Smith's cough," and so it proved.
>
> Another gentleman unconsciously laughed and the operator at the other end immediately detected his presence. Any person can use a telephone as readily as an ordinary speaking-tube. It requires no practice, and will be found so convenient that soon scarcely any business house in the city will be found without one. As a means of communication between the dockyard and the city; the waterworks and the city hall; the gas works and the office in town, a telephone would be found invaluable. It would save horse-hire, shoe leather and man-wear. How far can the human voice be projected by aid of the telephone? The longest distance to which a test has been applied is 2,000 miles, and a conversation was successfully carried on. The telephone is an instrument of great possibilities.[5]

In 1880 a number of events helped launch the telephone industry as a serious economic undertaking in British Columbia. McMicking's new-found career as telephone agent landed him in hot water with his employer, for it seems that he had used company funds to purchase his telephones. The general superintendent of Dominion Government Telegraph ordered an audit of McMicking's books, and at the end of March, 1880 the company dismissed him. McMicking was not too concerned as he was already meeting with a number of Victoria businessmen to establish the Victoria and Esquimalt Telephone Company.[6]

The major investor in the new company was Edgar Crow Baker. Baker had met McMicking through the Masonic lodge and had persuaded other Masons to help "bail out" McMicking financially when he ran afoul of the telegraph company. Baker was an influential figure in the Victoria business community, and it was his financial support and business ties that initially made the telephone a viable enterprise in British Columbia. Baker had settled in Victoria after a career in the British Navy, becoming a notary public, a real estate promoter, a member of parliament, and finally senator. He had solid links with legislators in Victoria and Ottawa, which aided the telephone company in a number of its later disputes with municipalities.[7]

On May 8, 1880, an act of the British Columbia legislature established the Victoria and Esquimalt Telephone Company. The business was incorporated with $5,000 in capital, at $100 per share, and empowered to operate telephone service in Victoria and Esquimalt for ten years. The enabling legislation set rates at a maximum of $5 a month per subscriber, with additional charges permitted for the use of telephones by non-subscribers and for the delivery of messages. It suited the telephone company to have its corporation written into legislation because it needed access to public lands and streets. This far-reaching permission for access was possible only through an act of the legislature. Though the company did not immediately anticipate problems with the municipal government (the city mayor was a Mason and a close friend of Baker's), it sought the legislation as insurance for the future.[8]

By 1883 Baker held 26 out of 55 shares in the Victoria and Esquimalt Telephone Company. He took an active role in the running of the company and eventually came to control 70 per cent of its shares. Baker stayed on as manager after the company was sold to Vancouver interests in 1899.[9]

The first telephones on the mainland appeared shortly after the Victoria and Esquimalt Telephone Company was formed. As with early developments on the island, the first sets functioned simply as intercoms. McMicking sold a set of telephones to missionary William Duncan, who set up communications between the sawmill and the general store at the site of his mission, the Indian village of Metlakatla, near the present site of Prince Rupert. Prince Rupert holds the distinction of not only being the first area to use telephones on the mainland, but of being today the only municipality in British Columbia possessing its own telephone company.

Another early customer on the mainland was Andrew Onderdonk, a contractor for the Canadian Pacific Railway. Six years before the placing of the last spike and the completion of the CPR, Onderdonk commissioned McMicking in 1880 to install a fifteen-mile telephone line between his home in Yale to the main railway tunnel along the Fraser River. Onderdonk was quickly converted to the use of telephones. A few years later he signed a contract with a meat contractor who serviced the railway crews. The contract included the stipulation that a telephone connection would be established between the slaughterhouse and the proposed railway terminus at Port Moody.[10]

Encouraged by Onderdonk's contract requiring the telephone link from the slaughterhouse to the railhead, two real estate speculators in Port Moody formed a telephone company. They received permission from the city of New Westminster to erect telephone poles from the Royal City to Port Moody, and by November, 1883 there was telephone service between the two cities. They lobbied for, and received in February, 1884, an act of the legislature granting a charter for the already-operating New Westminster and Port Moody Telephone Company. The act was similar to the act which established the Victoria and Esquimalt Telephone Company, right down to rates.[11]

Almost from the beginning, the New Westminster and Port Moody Telephone Company was in financial difficulties. Though the capital required to build a telephone network was relatively small by today's standards, a network had to be constructed and maintained. As a result the company ran up large debts with the Bank of British Columbia and a local entrepreneur, Joseph C. Armstrong. When the CPR decided to move the terminus of the railway to the coast, the Port Moody real estate market crashed and the company went bankrupt.

Armstrong and the manager of the Bank of British Columbia took possession of the telephone company. They saw that the only way to protect their holdings in the company was to upgrade and expand its facilities. They therefore extended telephone facilities to the coastal village of Granville. On April 6, 1886, the same day that Granville was incorporated as the City of Vancouver, the legislature granted an amendment to the New Westminster and Port Moody Telephone Company's act of incorporation. The amendment changed the name of the mainland telephone company to the New Westminster and Burrard Inlet Telephone Company and allowed it to extend its interests to Vancouver. The company's capital base was also increased to $100,000.[12]

From its new base in what would emerge as the booming commercial capital of the province, the telephone company underwent unprecedented expansion and consolidation for the next twenty years. In 1886 Dr. James Matthew Lefevre, a surgeon with the CPR and a Vancouver businessman, purchased an interest in the company. A few years later he became a member of the board of directors, where he was to play a major role in directing the company.

By the 1890s the expanding industry had attracted numerous eager investors and there was a proliferation of telephone companies in the interior of the province. In 1891 the New Westminster and Burrard Inlet Telephone Company launched a subsidiary, the Vernon and Nelson Telephone Company, to provide telephone service in the booming mining centre of the southern interior. The Vernon and Nelson Telephone Company was to face and survive a decade of competition. To finance its battle for monopoly, or at least dominance, the directors of the New Westminster and Burrard Inlet Telephone Company turned, like many a Canadian enterprise in this period, to Britain. In 1898 the company was sold to the Yorkshire Guarantee and Securities Company of Huddersfield, England. Yorkshire already had a number of interests in British Columbia which were supervised by its Vancouver resident manager, William Farrell. The British company incorporated their interests in the young Canadian company by

Setting a New Westminster & Burrard Inlet Co. pole in the 1890s near the current site of the seabus station in Vancouver.

establishing a new British company, British Columbia Telephones Limited. By 1901 BC Telephones had controlling interests in telephone companies in Vancouver, Victoria, New Westminster, Nanaimo and Kamloops, as well as in the Slocan, Boundary, and Kootenay Lakes areas.[13]

In 1902 control over BC Telephones returned to Vancouver when Dr. Lefevre, together with an English director of BC Telephones, purchased a controlling share in the company. Dr. Lefevre used the Vernon and Nelson Telephone Company, rather than the parent company, to amalgamate and consolidate his interests. The interior company held a perpetual charter, while the Vancouver company's charter was due for extension. In 1904 the company became the British Columbia Telephone Company Ltd. By 1906 Dr. Lefevre and William Farrell were its main shareholders.[14]

As the telephone industry expanded and underwent a number of technological developments a new skilled work force developed. The earliest telephones did not need an operator — at least not in the way that the telegraph needed an operator. Any person could speak or listen for themselves. As the first sets were hooked up and directly linked to each other by wire, there was not even the need for a switchboard to connect one subscriber to another. But developers quickly recognized that if the telephone was going to sell widely, subscribers would want to be able to talk to other subscribers in the area.

The introduction of switchboards marked the first significant change in the telephone system and permitted subscribers to communicate with each other. But a switchboard, unlike the earlier pairs of telephones, required an operator. In contrast with the telegraph operator, the telephone operator did not require a great many special skills, needing only to work quickly, efficiently and inexpensively.

With few subscribers, the operator was not very busy, and the job included other duties. Child labour was the cheapest form of labour, so boys were employed as the first operators. The decision was logical since boys were already working in the emerging communications business, frequently as messengers with the telegraph companies. In the early years telephone operators were required to run messages and so many of these boys worked as operator/messengers. One of the first boy operators in Victoria was Ernest Halliday, who worked for the Victoria and Esquimalt Telephone Company from 1881 to 1883. Starting less than a year after the company was formed, Halliday recalled that during his tenure all operators were boys. "There were 90 phones when I joined and 125 when I left the service. We boys had to sweep the floor and do the collecting as well." Baker himself supervised the collections and docked the boys for shortages out of their meagre pay of $10 per month.[15]

Later in 1881 the company, at the request of the city council, initiated a 24-hour telephone service. The late night service was primarily for emergencies, of which fire alarms were the most common. The boys on the night shift usually worked from 11 pm to 7 am, although a shift from 9 pm to 7 am was not unusual. There was no shift differential, but later that year the boys got a wage increase, raising their salary to $15 a month. A few years later this was increased to $20, where it stayed until operators organized and took collective action in the early years of this century.

The first telephone exchanges were not located in separate buildings but were rather switchboards in stores. Frequently the operator who staffed the switchboard was a relative of the storeowner. Sometimes the operator would work as store clerk as well as telephone operator. In Vancouver, for instance, the switchboard was located in Tilley's Bookstore, and the operator and manager of the Vancouver office was Charlie Tilley, son of the store owner. Evening business on the switchboard was usually slack, and Charlie amused himself and the subscribers by arranging nightly musical concerts broadcast over the telephone lines from the office. There were only about a hundred telephones in Vancouver, and the telephone office became a nightly rendezvous for young people who enjoyed serenading the subscribers, even answering song requests. The concerts always ended by 10:30 with the singing of "God Save the Queen." According to J. H. King, who worked as a boy operator at the Vancouver exchange, there were only himself and Tilley working as operators at the exchange in 1888. Yet they offered 24-hour service. King worked from 7 am until 9 am in the morning and from 5 pm until 10 pm at night. Tilley was on duty the rest of the time. At night Tilley slept in the telephone office with a bell near his bed to summon him to the switchboard when someone called.[16]

Frequently the people setting up telephone agencies or offices were from the telegraph industry. In Nanaimo a telegraph operator tried to copy McMicking's success in Victoria by launching a telephone company in the 1880s. The operator, J. A. Callaghan, successfully involved local businessmen in the project. They purchased equipment from Victoria and established their own telephone network, even holding a bee to gather volunteer labour to set up the poles. A local store became the home of the exchange, with the merchant and his wife as operators.[17]

In the late 1880s telephone companies throughout North America began to employ women operators. Management felt that boys tended to be rude to subscribers, played practical jokes and needed constant supervision. Young women, on the other hand, worked hard and continuously, and even worked without pay for several months while they trained to become telephone operators. Women operators were paid the same wage as the boys. From the 1890s onward, telephone operating became an almost exclusively female job. In 1891 only 4.4 per cent of the Canadian work force was women, the majority of whom were employed in domestic service and factories. Only a small number of professions were open to women and compared to most job possibilities, telephone work was clean, interesting and prestigious; indeed it was tremendously exciting. The telephone was still a wonder and people would stroll to the telephone office to watch operators talk to strangers miles away. There was responsibility attached to the job, since telephones were used to report fires, accidents, and other emergencies; the speed, quick thinking and efficiency of an operator was often credited with saving life and property.

The men who constructed the physical plant of the telephone network, set the poles, strung the wire, installed the switchboards and kept everything in working order, came to the industry from the telegraphs, the railways and even from sailing ships. The closest related trade was telegraph line work, and not only the telephone companies but the street railway and electric street light companies sought linemen with experience. There were no telephone experts as such, but skills from other trades were often valuable in the new trade. George P. Kelly, one of the early telephone construction foremen in Victoria, had learned to handle spars and to splice from his experience at sea.[18]

The first telephone lines were open or bare wires made of iron. They were hung casually on whatever was convenient, including trees, hedges and buildings. Such haphazard connections tended to malfunction more often than not, and the telephone companies began to erect a network of telephone poles to overcome this unreliability. Most construction work was done with little or no mechanical assistance, a team of horses being the extent of the aid that a construction crew could expect. As telephone service increased, cross-arms were added to the poles to accommodate more lines. However, the introduction of street railways and electric street lights created interference which caused noise on the telephone lines. In Victoria the company erected 80-foot poles, rising 70 feet above the ground, in an attempt to eliminate the interference. In the 1890s a further innovation saw the conversion of the entire system from ground return to metallic ground circuits which helped to eliminate the interference.[19]

Wire work with the telephone company was hard and dangerous

A heavy gang construction crew for Vernon & Nelson Telephone Co., turn of the century.

and involved long hours. The proliferation of poles and electric wires across each city created a major safety problem for linemen. Aside from the obvious danger of climbing a pole and working twenty feet or more above the ground, there were few regulations requiring separation of telephone lines from "hot" wires. On more than one occasion sagging telephone wires touched a hot wire and electrocuted a man working on connecting line miles away.

Most of the men who worked on the physical plant of the telephone company were hired for a particular construction project and laid off when it was completed. But there was also a small permanent craft work force. They were responsible for maintaining switchboards and other equipment at the exchanges, as well as for installing and repairing subscribers' telephones and for line work.

At the end of the nineteenth century subscribers were counted in the hundreds rather than thousands, even in the main cities of Vancouver, Victoria, and Nanaimo. One wire chief or foreman and two linemen were frequently the only permanent plant staff. When work was slack, these workers helped collect delinquent accounts and occasionally even staffed the switchboards if telephone traffic was heavy or an operator was absent. Because the companies employed both permanent staff and temporary crews, it is difficult to estimate the number of men who worked for a telephone company. Also most companies were expanding and frequently hired new staff to build extensions or to upgrade the system through conversions from iron to copper wire or the changeover from ground return to metallic ground.[20]

Linemen and electrical workers in British Columbia trace their organization to the turn of the century, when a speaker from the American-based International Brotherhood of Electrical Workers (IBEW) was invited to address a meeting of Vancouver electrical workers. A number of other craft unions organizing skilled workers were already established in the city. Although compared to other crafts the electrical trades were relatively new, they had already an interesting organizing background. As early as the 1870s telegraph workers began to organize in both Canada and the United States, and telegraphic linemen and operators helped organize one of North America's earliest industrial unions, the Knights of Labour. In 1883 telegraph workers belonging to the Knights of Labour declared a general strike against the telegraph companies which ended in failure and broke the union. A year later a new organization—the United Order of Linemen—was started in the West. It was a secret society restricted to linemen. The secrecy was thought necessary after the telegraphers' experience in the Knights of Labour debacle, in which linemen who supported labour organizing were blacklisted. As a secret society the United Order refused to link up with other labour organizations. Many linemen refused to join the organization because they were leery of secret societies. The nature of line work in the 1880s and 1890s was such that a large number of linemen, called "boomers," moved endlessly from job to job across the continent; they did not stay in one place long enough to establish a lodge of the United Order of Linemen. By 1890 the Order had disappeared, except for isolated lodges on the Pacific Coast. There is no record of a lodge in Vancouver or whether linemen involved in the Knights of Labour or the United Order of Linemen travelled to Vancouver. But in view of the nature of their work at the time, it is entirely possible.[21]

A new electrical trades organization was born in 1890, when St. Louis hosted an Exposition of Electrical Wonders. The exposition brought electrical workers from all over the United States to stay in St. Louis for a number of years while they built the displays. It was an organizer's dream. Many of the electrical workers were experienced union men, former members of either the Knights of Labour or the United Order of Linemen. The American Federation of Labour (AFL) sent an organizer to St. Louis and a new organization was formed almost immediately—the Wiremen's and Linemen's Union, Local 5221, under direct charter from the AFL.[22]

As important as it was to establish this first local, the workers realized the need for a national organization to have an effective voice in the rapidly-growing industry. When electrical workers left St. Louis at the close of the exposition, the union encouraged them to form locals in their home cities.

The AFL helped the Wiremen's and Linemen's Union establish contacts in the electrical trades in cities throughout the United States. In the fall of 1891 a call went out from St. Louis announcing a national convention of electrical workers. Small though it was, with only ten delegates from eight local unions, this convention launched the National Brotherhood of Electrical Workers, an AFL-affiliated union with jurisdiction over all electrical work.

The growth of the Brotherhood "followed the wire," with boomer linemen and wiremen bringing the union to each new city or job site. In these early years each worker was in practice a shop steward, ensuring that new employees joined the union and enforcing union-determined safety requirements and conditions of work. Grievances tended to be resolved at the job site, with workers downing tools until an issue had been settled.

The union also played a crucial educational role. Electrical workers learned the skills of their trade from their more experienced co-workers. The Brotherhood's constitution included an apprenticeship program which the union saw as controlling the number of tradesmen and maintaining a high standard of skill. Apprenticeships were restricted to men in their teens or early twenties. The union felt that it was more difficult to teach older men, who would not easily accept the strict discipline of the four-year tutelage. The union also feared that older men would tend to learn enough of the skill in a year or two to work in the trade without bothering to complete the apprenticeship.

The Brotherhood's growth in its early years was remarkable; by 1899 it had become a national organization in fact as well as in name. A convention that year decided to extend the organization of electrical workers into Canada and changed its name to the International Brotherhood of Electrical Workers.[23]

The Brotherhood's decision to expand into Canada coincided with a boom in British Columbia's resource-based economy. The completion of the CPR had created a period of unprecedented growth. While the forest industry at first experienced a downturn with the ending of railway construction, it quickly recovered by responding to the opening of the prairie markets and their demand for lumber. In 1898 the CPR purchased rail lines in the Kootenay region, opening the mineral and metallurgic empire in the southern interior.

The early unions in the province developed during this boom. By the turn of the century there were nineteen locals of building trade unions. Among the earliest campaigns of the labour movement was the demand for the shortening of the work day to nine hours. The campaign brought labour organizations around the province together and created labour councils in many cities. While union demands were directed at individual employers, the campaign was also political, with the unions seeking an eight-hour legislated work day. The British Columbian labour movement from inception was political, recognizing that for labour to make headway in a province dominated by big business ownership of the mines, railways and forests, labour needed political as well as economic action.

At the turn of the century a combination of active labour organization and the growth in the electrical trades in Vancouver saw a group of linemen and telephone workers meet in the fall of 1901 and decide to organize a branch of the IBEW. On November 6, 1901, they were granted a charter for IBEW Local 213. A few months later workers organized a New Westminster branch and soon after that, in March, 1902, Victoria Local 230 was formed.[24]

The new locals held monthly meetings which included a one-hour "lecture on practical electric subjects in three branches, light, telephone, and inside wiring." The union elected its own leadership and adopted, at local meetings, demands to be presented to electrical contractors and the telephone company. The membership was in constant flux, with travelling wiremen and boomer linemen moving from local to local in search of jobs. Vancouver Local 213 arranged with certain hotels to provide newly-arrived electrical workers who possessed travel cards from other IBEW locals with three meals and a night's lodging on the credit of the local.

In the fall of 1902 the women operators at the New Westminster and Burrard Inlet Telephone Company joined Local 213 as a women's auxiliary. The women's auxiliary was in fact a sub-local of Local 213. The term "auxiliary" usually refers to an organization of the wives of workers in a particular union, not a sub-local. The auxiliary conducted its own meetings, elected its officers and operated as an independent local, with two male delegates from Local 213 attending meetings for purposes of liaison. The auxiliary was one of the earliest successful efforts to organize women workers in Vancouver.

When the IBEW was first formed, it was exclusively a craft union of wiremen and linemen, and the issue of admitting women did not arise. As the union started to organize telephone workers, female workers joined locals of the union. In 1893 the union amended its constitution to exclude women, but this met with strong opposition and the amendment was rescinded. Though not formally excluded from the union, women were not actively sought as members. It appears that the enthusiasm with which women were signed up and the nature of the organizational form that their membership took was determined by individual locals, at least up to 1911. Thereafter a separate operators', and by extension women's, department was organized.[25]

There were many obstacles which made it difficult for women to organize. Women were a tiny minority within the workforce. A large proportion of the women in wage labour were in isolated jobs, such as domestic service. Women workers also tended to withdraw from the labour market after marriage or when they had children. Finally, many labour organizations were hostile and regarded women's and children's labour as a threat to their members, undercutting their wages and taking their jobs. For these organizations the simplest solution was to ban women and children from the workplace. In spite of this general trend, the Vancouver IBEW adopted a relatively progressive policy towards their "electrical working sisters."

The operators derived obvious advantages from the union. Telephone work was becoming more regulated. Individual operators had little say in the planning and conduct of their work. Gone were the days when city operators could work in a small store, with a leisurely work pace and a sociable atmosphere. With the increase in workload came increases in staff, work speed-ups, management regulation, and wider distinctions between the status of operators and managers. In Victoria, for example, McMicking forced operators to sign a copy of the following regulations when they were hired:

> Operators while on duty must sink every other consideration to expedite the work for which they are employed.
>
> The company, realizing that the success of the system depends very largely upon the method and zeal with which the work at the central office is carried on, and that all contemporary efforts are in vain if confusion and unfaithfulness are to be found there, direct the attention of all employees to a careful study of the following rules, which will form the basis of contract between the company and its employees in the operating room.
>
> 1st. All operators are required to be in attendance promptly at the time assigned for duty, and when specially needed are expected to remain as long as necessary.
>
> 2nd. Conversation necessary with subscribers must be carried on in a polite tone, and prompt and patient attention given to their wants. Any semblance of incivility, pertness or fault-finding will not be tolerated.

Operators at one of Vancouver's earliest telephone exchanges, 1890s.

3rd. Operators will not speak for the company on matters outside the ordinary routine of switching duties, but will receive and carefully note any requests made by subscribers or others, for reference to the proper officer.

4th. While on duty operators must sit at the switchboard with head telephone properly attached, having both hands free to work the switches.

5th. As successful switching cannot be accomplished with a board encumbered with cords, it will be the duty of each to strive at keeping the board as clear as business will permit, while exercising all diligence in discovering how to be most useful in aiding business dispatch.

6th. As every instrument in circuit in the office will transmit in common any sound produced in the office, the work should and must be carried on quietly and with due regard for the convenience of patrons. Loud talking or laughing must be avoided.

7th. The use of the wires by employees is restricted to the business of the company. Idle or unnecessary conversation in the office or through the instruments will not be permitted.

8th. Engaging in any outside recreation or work, reading books or papers, or smoking in the office is strictly forbidden during hours of duty.

9th. Loitering in the operating room by anyone, whether connected with the company or not, is strictly forbidden; and no one but the operator or an officer of the company is to have access to the switches or switchboard instrument.

10th. Operators will do well to bear in mind that they occupy an important and confidential position, where they may be capable, by close application to duty and integrity of purpose, of doing much to advance the interests of the company and the community in general, and where through failure to apprehend duty they may retard such interests in common without effort.

11th. Any knowledge of the business of the company or of patrons gained by reason of the position held as operator must not be disclosed under any pretence whatever.

12th. Instruments and all machinery of the company must be handled with care. Every section is more or less an integral part of the whole, so that an apparently trifling defect may prove a serious inconvenience to patrons and a corresponding loss of status, if nothing more, to the company. All damage to instruments, being the result of carelessness or culpable indifference, will be charged against the person responsible for such damage.

13th. Prudence at the switchboard more than in any other position is indispensably necessary, hence the exercise of this qualification chiefly will recommend an operator to a continuance in service.

14th. Laying aside all private business, intercourse with friends, or other ulterior interests during office hours, operators will devote their whole attention to the business of the company.

> 15th. A wilful violation or disregard of any of the foregoing rules will be deemed to be tantamount to a request to be relieved from duty, and the engagement will be terminated accordingly, with or without notice, in the discretion of the proper officer.[26]

The document covered the company's concerns, but offered operators no protection or rights. As work in the industry became regulated and codified in written contract, the workers moved towards collective action in order to ensure their rights. A union with the power to negotiate a contract which set out working conditions, wages and the rights of labour was necessary to protect both operators and craft workers from arbitrary decisions by management.

Within a year of the founding of Local 213 most electrical workers in Vancouver had joined the organization. But management still did not recognize the union as a body which bargained for its members. The New Westminster and Burrard Inlet Telephone Company even then was one of the largest employers of electrical workers in the city. In September, 1902, more than a dozen linemen working for the telephone company objected when management instituted new operating procedures and insisted that the men provide their own tools. Differences came to a head on September 16, when the linemen walked off the job. The general superintendent of the company, H. W. Kent, circulated a message to other telephone companies in an attempt to blacklist the strikers. The strike continued until William Farrell, managing director of the company, agreed to the workers' demands. The first telephone workers' strike in British Columbia had been waged and won. The resulting agreement covered company recognition of the IBEW, reinstatement of the striking workers and a promise of no discrimination against them, a promise to furnish all tools except those electrical workers customarily owned themselves, and a notification which was sent to all other telephone companies that the dispute had been settled. A further demand for the appointment of an arbitration board, with one nominee from management, one from the union, and a third acceptable to the first two parties, which would be charged with settling further disputes, was put aside for future consideration.[27]

It seemed that the union had won a great victory when it achieved formal recognition. But the ease with which the company had conceded union recognition indicated that the company and the union did not have the same understanding of the term. It soon became apparent that the company recognized nothing more than that a large number of its employees were members of the IBEW. In the company's view "recognition" did not imply that the union had any right to bargain with the company in the name of its members.

This misunderstanding between company and union on an issue as crucial as recognition can only be understood in the context of the period. At the turn of the century industrial relations bore little resemblance to the voluminous field of today. The Dominion Department of Labour, for example, was only established in 1900. Written contracts were rare. Those that did exist were usually very brief, on one or two pieces of paper, and dealt almost exclusively with wage rates. More often than not, a contract was an employer's tool, such as McMicking's contract, to ensure that the employee had read and knew the regulations of the job. Certification of unions did not exist and union recognition formally occurred only when it was written into a contract. Even then recognition lapsed with the expiry of the contract. At every new bargaining session the union had to battle management anew for the right to speak for the employees.

To the workers at New Westminster and Burrard Inlet Telephone Company the recognition of their union in September was a tremendous victory. Over the next few months the separate meetings of the operators' auxiliary and the craft workers in Local 213 worked out demands for future negotiations between the union and the company. The operators' main demands were increased wages, sick leave with pay, and an end to the company's use of unpaid trainees. The craft workers wanted parity with Seattle telephone workers, who had just won the eight-hour day at a wage of $3.25 compared to their nine-hour day at $3.00. The comparison with Seattle was logical. Generally wages and the cost of living were higher in the west than in the east. As well, the closest IBEW local outside of the province was Seattle. By November the union had drawn up a contract, and on the 25th the draft was submitted to the company with a request for a reply by the following day. In his reply, addressed to the union representative as an individual and pointedly ignoring the union, general superintendent Kent told workers that the "document" could not be dealt with until the return of William Farrell and Dr. Lefevre, who were both away in Toronto.[28]

There was much speculation in the daily newspapers that the next move would be a strike by the union. A chief operator, asked by a *Daily Province* reporter how city residents would know if a telephone strike had been called, responded, "You ring up central, and if you don't get an answer you may know that the strike is on." Hours after the statement appeared in the press on November 26, there was no answer at central office. The "hello girls" and the electrical workers had gone on strike.[29]

The nature of telephone switching technology meant that the operators' strike completely paralyzed the city's telephone network. Almost all subscribers had to contact the central office's switchboard and be connected by an operator. The linemen's strike a few months earlier had not disrupted telephone service, but the operators' strike paralyzed city telephone service and became an immediate and pressing public issue.

General superintendent Kent, the company's spokesperson throughout the strike, charged the union with making impossible demands and with giving the company almost no notice before the strike. Worse still in his eyes, the workers had presented their demands when the two key directors, Farrell and Lefevre, were away. Kent further argued that the operators had been forced out on strike by the linemen's union. He told the press that the operators had informed him "individually in my office that they had no grievances" and that "they were satisfied with the way they had been and were being treated, but as members of the union they had to go out on strike."[30]

Kent's charges infuriated the workers. They felt the company was trying to deny them the recognition they had won in September, and as proof of this charge the union pointed to Kent's letter, which had made no reference to the union. The operators were particularly disturbed by Kent's suggestion that they were responding to pressure from the men. An auxiliary official told the press that the women had in fact considered striking two months earlier, shortly after the auxiliary had been organized, but the men persuaded them to wait. Other operators also detailed grievances to the city newspapers. Aside from their low pay, they were particularly annoyed at having to pay the salaries of replacements if they were ill. They protested against the use of unpaid trainees, referring to the case of one woman who had been operating for nine months before she was put on salary. Also the operators were supposed to have one Saturday afternoon in six as a holiday, but the company rarely gave them this half-day because of a constant shortage of operators. Finally the operators claimed that the company discriminated against operators active in the union.[31]

Two days after the strike began in Vancouver, New Westminster operators went out, closing down the New Westminster central office. A few days later linemen in Victoria, members of Local 230, struck the Victoria and Esquimalt Telephone Company in support of their Vancouver sisters and brothers in Local 213. The Victoria operators were not members of the union and did not join the strike.[32]

At this point the strike took an unusual turn. From the outset the business community was the sector most affected by the strike.

There were about 1,000 telephones in Vancouver and most were rented by business concerns and their owners. At the suggestion of the *Daily Province*, a group of prominent businessmen including W. H. Malkin, president of the Board of Trade, and H. T. Lockyer, manager of the Hudson's Bay Company, met to discuss how telephone service could be restored pending final settlement of the dispute. They came up with a novel scheme: a group of businessmen headed by Lockyer and Malkin would take control of the telephone company. This would permit the resumption of telephone service until the return of the company directors.

The union recommended to its members that they return to work under the supervision of the businessmen's committee, but the company refused to accept the scheme and it fell through. The businessmen's committee laid the blame for the failure squarely on the telephone company and published the following letter on the front page of the *Daily Province* on December 1, 1902:

> We, the undersigned committee of businessmen, who on Friday afternoon last waited on the manager of the telephone company, with a view of making some arrangements whereby the terrible inconvenience being suffered by the business community by the cessation of the operation of the telephone system might be alleviated, think it advisable that a full statement of facts should be given to the public as to our negotiations.
>
> COMPANY AT FIRST CONSENTED
>
> As a result of the interview above referred to, we obtained from the manager of the telephone company his unqualified consent and approval, that he would be willing to enter into any agreement whereby the system could be temporarily operated, pending the adjustment of the differences between the company and employees. The manager stated most emphatically that he was prepared to place himself and his system entirely in the hands of the committee, and was willing to spare no expense provided we could arrange for the temporary working of the system.
>
> STRIKERS' REASONABLE STAND
>
> The employees' committee met our suggestions in a very commendable spirit, and stated that they would give us an answer the following day; this was done at noon on Saturday, when the chairman of the employees' committee stated that out of regard to the wishes of the businessmen they would operate the system for a period of a week under the superintendency of our committee, so that pending negotiations with the officials of the telephone company the business interests of the city would not suffer.
>
> On Saturday evening our committee again met the employees' committee, when details for the temporary operation of the system were arranged. The employees stipulated that our committee would undertake that by their action in this respect they would not be prejudicing their position with the telephone company, and that the

telephone company on the other hand would undertake not to use the week of temporary operation for the purpose of strengthening their position by engagement of outside help during that period or by any other means.

COMPANY BACKED DOWN

Negotiations with the employees' committee having been satisfactorily concluded, our committee waited on the manager of the telephone company to obtain an agreement from him in writing along the lines of what he had already verbally consented to do, when, however, it was stated to the manager that such an agreement would have to be made in writing and supplemented by a bond of $5,000.00, executed on behalf of his company as a guarantee of good faith, he flatly refused to negotiate further.

Subsequently it was mentioned to the management that one of the conditions of the employees returning to work under the direction of our committee would be that we furnish exclusively the operators for the week, but that they be all employed under his sole management and direction, this arrangement also, he took exception to and finally it was agreed to leave the matter in abeyance until noon Sunday, when he was to give a definite answer.

COMPANY'S REGRETTABLE ACTION

At 1:45 pm yesterday [Sunday] the matter was definitely concluded by a statement from the manager of the telephone company to the effect that he would proceed no further in the matter, which answer was immediately conveyed to the secretary of the employees' committee.

The committee cannot help expressing regret that their labours have not terminated more satisfactorily, particularly so as the arrangements entered into between the employees' committee and ourselves would have resulted in the telephone system being now in active operation.

W. T. Lockyer
W. H. Malkin
F. R. Stewart
W. J. McMillan
W. H. Ker.[33]

The frustration of the businessmen's attempt to find a solution meant that subsequent publicity was favorable to the union. From the beginning of the strike the press and the public had sympathized with the operators' demands for higher wages. When the company backed out of the businessmen's scheme, it became clear that the union could not be held responsible for the lack of telephone service. Even in 1902 public perception of a labour dispute was important, although economic power had played a significant role as well.

On the same weekend that the businessmen's committee was arbitrating between the company and the union, the company was pursuing its own solution to the dispute. The company had placed employment advertisements in the newspapers shortly after the strike began, and it recruited a few local women to try their hand at telephone operating. On Monday evening the central office reopened for service, staffed by 25 strikebreakers. Sleeping quarters were set up for them in the telephone office and their meals were sent in from a neighbourhood restaurant.

Over the next few days the weather intervened on behalf of the strikers. A splendid coastal rainstorm tangled lines and knocked down wires in most of the outlying areas, leaving only the telephones in the downtown business centre still working. The storm greatly increased the number of "troubles" (the company term for out of order equipment). Although the company managed to recruit a few strikebreaking linemen, the system continued to deteriorate throughout the strike.[34]

The city's central labour body, the Vancouver Trades and Labour Council, decided to rally support from the city's unions to aid the affiliated telephone workers. The council saw the company's response to the strike as an attempt to break a member union. After all the telephone workers had won union recognition and now the company was attempting to rescind it. The council placed a boycott on the restaurant which served food to the strikebreakers and organized rallies and fundraising activities.

The return of Farrell and Lefevre did not improve the situation. Farrell told the press that he was prepared to pay a wage comparable to other telephone companies in Canada, but the demand for parity with United States rates was not acceptable. He assured the press that he was willing to meet with any of his "former employees" at any time. The union and the press both interpreted Farrell's statement to mean that he was unwilling to meet with elected union representatives; however, when Farrell subsequently denied this, the way was opened to resume bargaining.[35]

In early December union and company officials met in an attempt to reach an agreement. Later three city aldermen were drawn into the talks when it appeared that they were getting nowhere. Predictably the two major issues were union recognition and the fate of the strikebreakers. The company insisted that the strikebreakers retain their jobs, while the union wanted them fired. Finally a compromise was reached: the union got its closed shop and strikebreakers who wanted to stay on had to take out membership in the union. The strikers won nearly all of their demands and signed a contract remarkable for the period.

The closed shop was the union's greatest achievement. It meant that all employees now had to join the union and that non-union staff could not be recruited to undermine the strength of the union in the workplace. The contract applied to "all electrical workers" including the operators, and exempted only the chief inspector, the general foreman, the chief day and night operators, and the very small office clerical staff.

Operators won an incremental pay scale, starting at $20 per month and reaching $30 after two years' experience. Trainees were to be placed on probation lasting no longer than ten days, after which they were to be paid at the $20 starting rate. The women were permitted up to three days' sick leave per month, with pay. The rotating Saturday afternoon holiday for operators was retained with a promise by the company to hire enough staff to allow the operators their time off. Craft workers won the eight-hour day and an immediate increase of $5 a month, which brought the journeyman's wage to $3.20 a day. In order to ensure that the company did not use cheaper apprentice labour to replace journeymen, a ratio of one apprentice for every two repairmen was written into the contract, along with the stipulation that apprentices could not install telephones. Finally the contract guaranteed that there would be no discrimination against strikers and gave company recognition to a union-appointed grievance committee. The life of the contract was to be one year from the date of its signing (December 15, 1902), with the provision that it would be renewed automatically after it expired unless one month's notice was given.[36]

The two-and-a-half-week 1902 strike was won because of a combination of factors. The telephone industry had already become an important service for business, with many firms relying heavily on this new form of communication. This meant that the strike wielded significant economic power. Also technology was such that operators were absolutely crucial to the functioning of the telephones. Furthermore the system was unreliable and could not function for even a short period without experienced repair staff. The bad weather helped speed up the normal deterioration of the network.

The union, and the "hello girls" in particular, had the sympathy of the public, who generally acknowledged that they were underpaid. Moreover their work had not yet become so regulated or impersonal that telephone subscribers felt indifferent to the voice from "central." The company had made a major public relations blunder in negotiations with the businessmen's committee, thereby alienating a group which included some of its most influential subscribers. Management recognized that they had to cut their losses in the face of this blunder. The directors, who had just retrieved control of the company from Britain, could not afford to have the strike drag on and undermine their new investment in what was now a locally-owned company. Finally, strike action was taken quickly, with little notice. The company had no time to recruit strikebreakers before the strikers walked out. The telephone company learned from its errors; it handled subsequent labour disputes more carefully. The union, on the other hand, gained a false sense of security from the closed shop provision in the contract. Worse still, it did not fully understand the unique set of circumstances which had allowed it to win a relatively easy victory.

Chapter 2 *The Employers' Offensive*

In the first years of this century the British Columbia labour movement grew rapidly in organization and militancy. The boom in mining, lumber, fishing and construction greatly increased the work force. Expansion also brought inflation, which cut into workers' wages. The new prosperity spurred the workers to organize and demand a share in the newly created wealth through better wages, shorter worktime, and safe working conditions. By 1903 the British Columbia labour movement had formed 216 local unions with a membership of more than 15,000 members. The focus of many strikes was the demand for improved conditions and union security through the closed shop. Important sectors of the economy were struck, including fisheries, mining and railways. Most strikes were not as successful as the telephone workers' strike of 1902.[1]

But workers were not the only ones to organize. By 1905 British Columbia employers had formed 26 associations, the majority of which were organized between 1902 and 1905. The fiasco of the 1902 telephone dispute—when leading businessmen had lined up in opposition to the struck telephone company in support of labour—was a major embarrassment to all employers. It was an experience they were not anxious to repeat. Significantly, shortly after the 1902 strike, William Farrell was elected president of the powerful British Columbia Employers' Association.

Many of the employers' organizations practised price fixing to maximize their profits while demand was high. As well, employers' organizations throughout the country launched a campaign to destroy unions and the closed shop. Unions, they argued, were conspiracies aimed at denying owners their rights to manage as they saw fit. The employers claimed that the closed shop was a denial of basic freedom which attacked the worker's right to work. Employers' journals argued, "to sell that labour to whomsoever he likes, wherever he likes, in such quantity as he likes, and at such rates as he likes, is a God-given privilege which should be the duty of . . . Parliament to protect . . . Any encroachment upon this privilege. . . can only be regarded as an unwarranted interference with individual and property rights." The employers regarded the union move to protect older workers through seniority clauses as unacceptable. Management would decide all questions of work assignment and security on the basis of merit. The attempt by unions to win established wage levels for all workers within a given job category were seen as detrimental, with employers contending that these practices removed the employee's incentive to work harder. The employers condemned unions as an outside influence causing discontent among an otherwise-satisfied work force. Unions, through the strike weapon, wielded anarchy and destruction on an otherwise orderly and just society.[2]

As a key leader in the employers' offensive, William Farrell was not about to accept the closed shop at British Columbia Telephone. Although the 1902 strike had won the closed shop, the union's position was far from secure. The closed shop applied only to the Vancouver operations of the company. Though BC Telephone now ran most of the telephone companies in the province, the union had not been able to extend its closed shop beyond Vancouver. In Nanaimo, Victoria, Kamloops and the southern interior, the exchanges were much smaller. Aside from Vancouver, only Victoria had a local of the IBEW. But the small telephone work force required in this city made it far easier for the company to replace union workers with strikebreakers, effectively preventing Local 230 from organizing the Victoria telephone exchange.

The unions regarded the closed shop as fundamental to sound labour relations. The company would contact the union when it wanted to hire new employees and the union would send the workers. If the union did not have the required workers, then the company could hire them directly, provided the new employees joined the union. The telephone company, however, ignored the union's hiring hall and instead hired employees directly. At first, as there was no shortage of jobs for craft workers, the union did not enforce its hiring hall procedure, although the union was diligent in signing up new craft employees.

The position of the telephone operators' auxiliary was precarious. As organized operators were the rare exception in the male-dominated labour movement, they were isolated within the labour movement itself. The rapid turnover among operators also prevented continuity in experienced leadership. Most operators were between the ages of 17 and 25. They rarely stayed longer than four years at the telephone company, usually leaving to get married or to have children. The few who stayed on became supervisors and chief operators, and often came to identify more with management than the union. The company selected and trained this ongoing wave of new operators, and the auxiliary was unable to keep up with the influx. The auxiliary did no educational work to make up for the inexperience of its new members or to explain the advantages of unions. Frequently the auxiliary allowed operators to work for years without becoming members of the union. The closed shop in practice never applied to the operators.

The craft workers were more successful in maintaining the closed shop. Their union possessed an education program which trained and examined apprentices. While the main purpose of the program was craft training, basic union education was included.

With the amalgamation of the provincial telephone interests into the dominant BC Telephone Company, the union felt it necessary to extend the closed shop throughout the company. Without an effective closed shop, the company could proceed with its desire to hire and train non-union telephone workers, which would enable the company to shut the union out of telephone work. Enforcement and extension of the closed shop was thus a matter of survival.

In April, 1905 the BC Telephone Company in Victoria locked out eight linemen, all of whom were members of IBEW Local 230. The strike of Local 230 in 1902 in solidarity with the Vancouver telephone workers had not won the closed shop in Victoria. Following the 1902 strike the local did not raise this issue because the company employed few craft workers. But in 1905 the company hired workers for a special construction project and all but one were union members. When the unionists discovered that one of the men on the project was not a union member, they refused to work with him. The company, anticipating a strike, locked out the union men.[3]

The union declared the BC Telephone Company in Victoria unfair and requested other locals to ensure that no unionists came to the island to work for BC Telephone. The company cancelled the construction project and maintained repair service with non-union labour. The union and the company finally reached a settlement four months later. The settlement revealed the union's weakness where only a few workers could maintain service—some union men were rehired and the closed shop was still not won.[4]

That year the Vancouver telephone workers suffered a serious loss with the death of Thomas Laurenson. Laurenson was a permanent employee with the telephone company and a central leader. He was not only shop steward for the telephone men, but also one of Local 213's delegates to the operators' auxiliary. His tragic death underlined the dangers in telephone work. He had been sent downtown to inspect a "trouble" and was electrocuted when he touched a hot wire. An investigation revealed that a builder working in the area had moved a telephone wire, inadvertently allowing it to touch a power line.[5]

Before his death, Laurenson had been drawing up new proposals for negotiations with the telephone company. He was also well aware of the need to strengthen the auxiliary. Since the 1902 strike the auxiliary had declined to the extent that by 1905 a majority of operators were not even union members. Local 213 assigned new delegates to work with the auxiliary, but it took time for them to win the same level of confidence among the operators that Laurenson enjoyed. The union remained concerned over the need to rebuild the auxiliary.

Despite the weakness of the auxiliary, the union decided to carry on with negotiations for a new agreement. The agreement, signed in January, 1906, did not include the operators, but the closed shop for the craft workers was maintained. This separation of the craft workers and operators proved to be a serious error. The union had allowed the company to undermine one of the most important achievements of the 1902 strike—recognition of all employees as a single bargaining unit.

General superintendent Kent had attempted to destroy the operators' union from its inception. He bitterly resented the 1902 victory of the "hello girls" and felt that a union of telephone operators was a direct challenge to his ability to manage. By 1905 he had initiated changes which succeeded in critically weakening the union. Aside from his major power to select and hire operators, Kent introduced a system of bonuses and special incentives which he awarded arbitrarily. He also promoted two key union activists to the newly created position of "monitor" or assistant "forewoman." The monitor jobs were payoffs in exchange for assistance in breaking the union. In early January Kent pressured 31 out of 37 operators into signing a statement declaring that they no longer wished to belong to the union. He had told the operators that if they did not sign this "yellow-dog" statement—the name given by the labour movement to this type of union busting tactic—they would have to leave their jobs. As a sweetener he offered them a new lounge and meeting place, as well as some improvements in working conditions. Kent's action backfired. A number of operators went to the union after signing the company letter and protested the coercive action. Within weeks the vast majority of operators were back in the union, along with new recruits who had turned to the union because they were angered by Kent's threat.[6]

This defeat did not deter Kent, and in February the company again pressured the operators to leave the union. This time Kent took each of the operators individually into his office and told them either to resign permanently from the union or to leave the company within 24 hours. The operators called an emergency meeting with the craft workers of Local 213, who expressed their full support for the operators' struggle. A joint grievance committee composed of both men and women unionists was sent to meet with Kent.[7]

The joint meeting had decided that the only way to stop the union-busting actions of the company was to gain clear recognition for the operators' union. To accomplish this, the union drew up a draft agreement which maintained existing working conditions and included a clause recognizing the union and establishing the closed shop. The company rejected this proposal on the grounds that the operator's union was a subsidiary of the IBEW and not a real union, and further, that the closed shop would require the company to fire non-union operators.[8]

On February 22, negotiations between the company and the operators' union broke down, and so 20 craft workers and 34 operators struck the telephone company. Unlike the situation in 1902, telephone service was not interrupted. The company had prepared for months for the strike. Kent had operator strikebreakers on hand, and his effort to undermine the operators' union had some effect in that he was able to persuade six operators to remain at work. As well, the company was able to import strikebreakers from Victoria and Nanaimo. McMicking, barely a year out of his own labour problems in Victoria, enthusiastically recruited scabs from his Victoria staff, though he complained to Kent that "we find more operators willing to go over, than we do Mothers ready to permit them to go . . ."[9]

The day after the strike started more than a dozen strikebreakers staffed the boards. The company recruited a few craft workers to keep the boards and switching equipment working. In spite of the strikebreakers the union was confident of its eventual victory. William Manning, president of the local, noted that the union was in a better financial situation and was better organized than it had been in 1902. But the workers made a grave error in that they saw the strike as a repetition of 1902. They were oblivious to the unique set of circumstances which had been on their side four years earlier and were slow to recognize the company's preparedness. The strikers' excessive confidence was bolstered early in the strike when they were able to win a few strikebreaking operators over to their side.[10]

At the beginning of the strike the company stated that it would be willing to allow M. J. Sullivan, vice-president for the Pacific coast region of the IBEW, to arbitrate the dispute. BC Telephone felt confident that no international officer of a craft union would sanction a strike for recognition of a non-craft "women's union." But Sullivan sent the following wire to Local 213 the day after the strike started: "Insist on the clause specifying that operators must belong to the union."

When the company learned of Sullivan's support for the operators, it refused to accept Sullivan as an arbitrator. Sullivan, however, arrived in Vancouver to help Local 213 negotiate with BC Telephone. After nearly a week, bargaining broke down; the company refused to grant recognition to the operators' union. Local 213 was as adamant in its refusal to abandon the operators.[11]

By the third week of the strike the company was again operating normally. The strikers were becoming frustrated with the strike's limited impact on the company and community. It was at this point that a number of telephone cables were slashed. The company made a much-publicized offer of $1,000 for information leading to the conviction of any vandal. BC Telephone suggested that it was the work of "linemen," as only telephone craft workers knew where major trunk lines were, and only they were capable of cutting the lines with such skill. In fact the company's own story tended to clear the linemen of involvement. The cuts were easily repaired and no moisture had seeped into the cable. This was fortunate for the company since moisture could have caused a short-circuit in the whole branch line. The telephone wires, enclosed in cable sheaths, were separated by paper insulation. Any moisture leaking into an opening in the cable would quickly spread, short-circuiting the entire branch line and causing a far more costly and time-consuming disruption of service. A lineman bent on sabotaging the telephone network was capable of wrecking the entire network.

Sabotage and slashings are endemic in the telephone industry. Significantly, this was the first time that the company had publicized the problem. From this point on, the issue found its way into every telephone strike. It helped to draw attention away from the real issues and to smear the credibility of the union. No amount of union denials or condemnation of this tactic could remove the suspicion that labour was responsible.

As March drew to a close, the network was severely damaged. The culprit, however, was a farmer who was blasting stumps near New Westminster. A piece from a blasted stump landed on the streetcar line, breaking the wire. The hot wire fell onto the telephone line, sending a charge down the line to the main Vancouver telephone exchange about ten miles away. Two switchboards in the central exchange burst into flames before the fire was smothered with sand. The switchboards had to be totally rewired, but little other damage was done. The firemen who had been called to the scene used the opportunity to discuss the dispute with the strikebreaking operators. Kent protested to the fire chief that the union firemen had shown a "singular lack of taste" in raising the issue of the strike "at a time of peril."[12]

The firemen had done their job, but in the spirit of union solidarity characteristic of the British Columbia labour movement, they did not pass up an opportunity to do a little union work. Solidarity was forthcoming from the city's labour movement and even from the occasional businessman. A large public meeting in support of the strike, organized by the Vancouver Trades and Labour Council near the end of March, featured a number of prominent labour leaders as well as L. D. Taylor, the business manager of the Vancouver newspaper, the *World,* who was later to become mayor of Vancouver. In his speech to the rally Taylor stated that since the telephone company was a "semi-public institution," it had a responsibility to act in the interest of the citizens. He condemned the union-busting tactics of the company and called for public ownership of the telephone company. John Davidson of the carpenters' union characterized the strike as a "challenge from the Employers' Association to organized labour." Davidson stated that the Employers' Association, through its president, William Farrell, was trying to discipline labour and return to the days of the open shop.[13]

The business community was no longer as sympathetic as it had been in 1902. The many strikes in the intervening years had done much to unite businessmen. The formation of the Employers' Association and the election of William Farrell as president closed their ranks even more. Because telephone service was maintained throughout the strike with little disruption, the strike did not directly affect as many businesses as in 1902.

As the strike dragged on, workers became more hostile toward the strikebreakers who had taken their jobs. Incidents, mostly fights between strikers and strikebreakers, resulted in the strikers being charged with intimidation or assault. The law usually came down heavily against the unionists.

An IBEW lineman working for BC Electric, Robert Emmett, was

fined $25—which amounted to over a week's wages—for calling two strikebreakers "scabs." The magistrate declared that the use of the word constituted grossly insulting language. Emmett in turn had laid assault charges against the two strikebreakers, who had attacked him. The magistrate dismissed the case against one of the attackers and fined the other a dollar.[14]

In another incident Arthur Tazzaman, a striking craft worker, attempted to take a photograph of a strikebreaker so that the union could ensure that the man would not be allowed in the union or on any unionized job site. The strikebreaker, spotting Tazzaman taking his picture, attacked him. The incident occurred outside the BC Telephone building on Seymour Street, with rather unfortunate timing. The chief of police happened along the street on his way to visit superintendent Kent. He waded into the fight and separated the two men. Kent, who had seen the fight from his office, rushed onto the street and smashed Tazzaman's camera. The police chief arrested Tazzaman, but released the strikebreaker on Kent's assurance that he would appear in court the following day. Tazzaman was charged with assault, but the case was eventually dismissed. Tazzaman and the union took Kent to court to recover the costs of the camera and won the case. The courts generally were not the least bit sympathetic to the strikers, and the union won its case only because Kent readily admitted that he "wilfully and on purpose" smashed the camera. The court gave him a token fine of a dollar and told him to replace the camera.[15]

In the spring, with no end to the strike in sight, the union attempted to establish a case in the small debt court for what today would be called an unfair labour practice. A leader of the operator's union, a woman named Carnahan, brought an action against the telephone company. She demanded one month's wages on the grounds that the company had ordered her to either quit the union or leave the company's employ. Carnahan claimed that she was dismissed for being a member of a union, but the magistrate found that she had remained at work after the company had issued the ultimatum and had only left work when the union had struck. The magistrate dismissed her claim. He did not consider relevant that management's threats had in fact provoked the strike. The case demonstrated major problems with the legal system as it applied to labour disputes. Though there was legal protection against dismissal for joining unions, discrimination was difficult to prove.[16]

There is no record of when the union officially called off the strike. As late as August, 1906, Local 213 passed a motion promising to "carry on the strike until we can come to an agreement." But by November a number of union linemen returned to work at BC Telephone without an agreement. The strike had ended in failure, and the telephone operators' union had been broken.[17]

The 1906 strike was one of a series of battles which broke out across the country over labour's demand for the closed shop. These confrontations spurred the federal Liberal government of Wilfrid Laurier to introduce labour legislation masterminded by the deputy minister of labour, William Lyon Mackenzie King. Entitled "An Act in the Prevention and Settlement of Strikes and Lockouts in Coal Mines and Industries Connected with Public Utilities," the law was generally known as the Industrial Disputes Investigation Act (IDIA).

The IDIA allowed either party in a dispute to request the appointment of a conciliation and investigation board to look into the dispute and file a report with the minister of labour. The report of the board was not binding, but the legislation banned strikes or lockouts until the board had completed its task and the minister was in receipt of the report. Many leaders of the Western Canadian labour movement opposed the legislation, saying the cooling-off period while a dispute was under investigation was a major constraint on the right to strike. Worse still, it allowed an employer time to recruit strikebreakers while going through the motions of "conciliation and investigation." As far as Western labour was concerned, the act was the result of collusion between employers and government.

The telephone workers had their first experience with the IDIA in 1911. By this time Local 213 had re-established itself among the craft workers at BC Telephone. The union had taken several years to recover from the defeat of 1906. They first organized the sub-contractors that BC Telephone used for most of its construction work. By 1910, Local 213 once again represented all craft workers at BC Telephone.

The workers were dissatisfied with craft wage rates at BC Telephone and the company's extensive use of apprentices and lesser-skilled workers at lower rates. The union sought parity with the better-paid Seattle telephone workers and with other unionized electrical workers in the city who worked for Western Power Company and BC Electric Railway. BC Telephone suggested that rates could be improved, if linemen were willing to work a longer day. But the union was adamant in preserving the eight-hour day. Negotiations then broke down. For months the local debated whether to submit their dispute to the IDIA. Finally in August it decided, on a close vote, to apply for an investigation board.[18]

At the suggestion of the telephone company, the IDIA-

A cable splicer, about 1915. Cable splicers were considered at the pinnacle of their craft in both the electrical and telephone industries.

Vancouver City Archives

appointed board confined its investigation to the issue of wages. The board, after hearing witnesses from both parties, sided with the workers and recommended that the company grant a 25-cent-per-day wage increase. The majority report pointed out that the work at the telephone company was "extra hazardous," requiring "young, strong, and active" workers, almost all of whom were under the age of 40. These workers were unable to afford life insurance because the risks involved in their work were reflected in exorbitant premium rates. In spite of the sympathetic response of the board, the union was annoyed that the board had limited itself to the issue of wages and ignored the pressing issue of the use of apprentices. Most members felt that the exercise had been a waste of time and decided to ignore the IDIA in future, particularly as the act placed constraints on the union's chief weapon, the strike.[19]

The company's use of apprentices challenged the very existence of the union. IBEW locals supplied most of the electrical workers in British Columbia. The union, through its apprenticeship program, trained and graduated skilled craft workers. As far as the union was concerned, there was no such thing as a "telephone man"—there were journeymen and apprentices. Electrical journeymen had at least three years' trade experience. They worked wherever work was available, whether that meant the electric railway, the power company, construction, or the telephone company. The loyalty of these men was to their union.

The telephone company held different views. It wanted to train its own people in specific telephone skills, including telephone installation and repair and switchboard maintenance. The company would thus be able to determine who was to be paid at the craft rate and who would be paid lower rates. Its goal was to create a workforce trained only in telephone work with non-transferable skills tied to the telephone company. Through this process the company hoped to free itself from the craft union.

For over a year the issue simmered, until in March, 1913, the electrical workers at BC Telephone went on strike over the issue. On March 1 the company had granted a 25-cent wage increase, bringing the daily journeyman wage to $4.25, but refused to negotiate the ratio of apprentices to journeymen. Tensions had risen in the months before the strike because some departments employed almost as many apprentices as journeymen.

The workers struck on March 14. The company immediately telegraphed Ottawa to request the appointment of a mediation board under the IDIA in the hope that the intervention of the board would force the strikers back to work. But the government instead assigned a Vancouver-based "Fair Wages Officer" to investigate the dispute. This mitigated against the company's strategy, as the assignment of an IDIA board would have forced the union back to work and given the company time to attempt to undermine the union.

The strike involved more than 300 telephone craft workers across the province and completely shut down the company's construction projects. This was the first province-wide strike and a major test of the unions' provincial strength against a by-now powerful company. Within ten days the strike was over. The men won a reduced ratio of apprentices to journeymen—one for every four journeymen—double time for overtime work, and a further across-the-board wage increase of 25 cents a day.[20]

While the craft workers had successfully opposed the company's attempt to control training, significant changes had taken place in the traffic or operator services department. The breaking of the telephone operators' union in 1906 opened the door to complete company dominance of the operators. The operators no longer possessed an organization that could resist the introduction of "scientific management."

The management of telephone companies were among the

earliest experimenters with this system of work control. "Scientific management," or Taylorism, was invented by Frederick Winslow Taylor and comprised three principles. First was a strict division between mental and physical tasks. The mental task of work planning was given to a management planning department which worked out manual tasks in the most minute detail. Secondly, workers were assigned a manual task which they repeated throughout the day. Finally, each worker was continuously monitored and timed to ensure the required speed and accuracy. The goal was to de-skill workers by giving management complete control over the planning and mental work. Under Taylorism workers did not need to know anything about the work they were doing; their work became an endless repetition of meaningless tasks. The minute timing and planning of each task allowed management to speed up work and increase production while it controlled every aspect of the work routine.[21]

Management's control over operators was nothing short of incredible. While it systematically degraded their work through application of Taylorism, the company propagated the myth of the heroic operators. Operators' work was indeed crucial at times. They deserved credit for alerting authorities to fire, accident, and other emergencies. But the company created a glamorous, exciting, professional image of the telephone operator to make the job more attractive to prospective employees. The image provided a vivid contrast to the low wages, strict discipline and patronizing attitude of the company toward the operators. In 1911 the telephone company initiated a monthly journal, *Telephone Talk,* which rarely appeared without some mention of operators. Tales of operator heroics, intelligence, and devotion to service paled the accomplishments of Joan of Arc by comparison. More regimented than most factories, an operator's work was nevertheless portrayed as a "prestigious" vocation for "young ladies." The company boasted that "Telephone operating has truly become a profession, and one of which any young lady may be proud. The school teacher, the stenographer and the telephone operator are on an equal footing, but of the three, the operator has no doubt taken the severest training for her work, and has the advantage of securing advancement to better positions and also of finding a position wherever she goes." The advertisement neglected to mention that the better position down the road was to become a supervisor of operators.[22]

By 1913, telephone operators at BC Telephone worked eight hours a day and had one day off each week. Almost all day work was split into two shifts of four hours each to maximize the number of operators on hand for the mid-morning and mid-afternoon

Vancouver City Archives

BC Tel's Fairview operators' school, about 1915.

calling peaks. Operators received wages ranging from $8.00 to $10.50 a week. Few received the higher wage, as it was reserved for supervisors. The wages were comparable to the $8.00 weekly wage of female factory workers or the $10.00 per week wages of waitresses. In Washington state, where the cost of living was slightly lower, the state had set a rate of $9.00 for a 44-hour week as the minimum for telephone workers. In Vancouver the Women's Council conservatively calculated that the minimum a woman needed to support herself was $7.50 a week. This minimum included $5.00 for board and room, 50 cents for car fare, 50 cents for laundry, and $1.50 a week for clothing. Operators were expected to wear gloves and hat to and from work and dress in a manner commensurate with their status as "young ladies."[23]

On an average, only 60 per cent of the many applicants were asked to report to the operators' training school. The company selected almost exclusively women from Canada, the United States and the British Isles. In 1914, for example, a random selection of acceptable applicants consisted of 35 Canadians, 33 Scots, 30 English, 6 Americans, 2 Irish and 3 Western Europeans.[24]

The criteria according to which the company selected trainees illustrate the fruits of scientific management. They included "intelligence, appearance, height, age (preferably between 17 and 25 years), experience, education and references." The company wanted women who could memorize standard phrases, which had to be rattled off with clear enunciation. As well, operators had to memorize the various positions of jacks and the procedures for processing calls. They were drilled in this work before winning full-time positions as operators. In the Seymour office, for example, there were 9,600 multiple jacks. An operator was expected to insert a plug from a caller into the slot of any of the numbers given to her without hesitation. She was also expected to monitor the boards in order to pull out the jack once the call was finished, to free the line for new calls. Height was an asset because it enabled operators to reach the numbers at the top of the board as well as work adjacent boards with as little effort and time lost as possible. A candidate for operator training had to pass a physical examination by a company physician. It took strength to sit rigid and alert with pounds of telephone apparatus suspended from the neck and head for eight hours a day.[25]

Until 1911, training of operators took place at the Seymour office. But heavy traffic at the office and an increasing number of trainees forced the company to open a special operators' training school with its own principal, teaching staff and elocution teacher. The school was an important component of scientific management at BC Telephone. The company preferred to hire young women with little or no work experience, claiming that they learned the fastest. They were also more willing than an experienced woman worker to tolerate the harsh discipline. The school trained the women to use specific phrases, each one carefully worked out so as to eliminate unnecessary words. Each action and statement they were taught had a specific time set for its completion. "Thorough training in the work, and absolute discipline" were the hallmarks of the operator school. Discipline was "the first point discussed with the students, and the last that they heard when assigned to positions as regular operators."[26]

The company's preoccupation with timing did not end when an operator completed her training. At least once a month each office counted the number of calls handled during the day. Operators were divided into various classifications, depending on the type of board and service, such as toll for long distance and flat rate for local city calls. "A" operators wre expected to handle 230 calls per hour from flat-rate telephones. During non-peak periods operators were often seated at every second position along a row of boards so that they could handle calls on two boards.[27]

The company soon recognized that more could be gained with the carrot than the stick, so speed-ups and efficiency drives were linked to incentives and contests. In 1913, for instance, the company introduced a "weighted service contest." This contest

Vancouver City Archives

A telephone operator, about 1915.

required management to monitor calls answered by operators and record extensive statistics for each office, including number of calls answered, number of misdirected calls, and amount of time spent per call. All these factors were given various weights with a perfect score being 100. The exchange which achieved the highest weighted rating in each month received a pennant with the company's shield on it, to be hung in the lounge for the month. In the first contest the modern new Bayview exchange won the pennant with 90 points, while the main Vancouver telephone factory on Seymour brought up the rear with a mere 69 points. The approach was ingenious. Under the guise of a game, the company extended its monitoring of operators while operators willingly increased their production and accuracy. All for the price of a pennant.[28]

As telephone service lost its status as a novelty and became a necessity, the number of operators rapidly increased. The company formed operator clubs which held meetings every two weeks in the operators' lounge. Minutes were kept of the meetings. Club meetings were essentially socials which consisted of a program of "piano and violin solos, of song and recitations," followed by "the serving of tea, cake, fruits and candies." The clubs were the paternalistic side of the company approach to the operators—management and operators socializing together "in good fellowship" and fun. Operators were treated as the beloved children in the stern telephone company family.[29]

The company formed a more exclusive club for supervisors in Vancouver. The object was to bring "chief operators and supervisors together to discuss points which come up in their daily work; to promote a better understanding with each other and to help in all ways to better the service; to impress on the operators that they are one large family, all working on the same lines, not one office against another."[30]

An important aspect of any family is the security it provides for the sick. In 1914 the company initiated a benefit fund and encouraged each operator to contribute 10 cents a month, which it would match. The company appointed supervisors to manage the fund and to see that help was given to operators "placed in difficult circumstances by illness."[31]

Despite all the talk of "working on the same lines," there remained a major difference between the company's goals and those of operators. In the final analysis the company was a business and a profitable one. In 1914 the company had made a profit of close to half a million dollars. Its objective was to make money by providing a service as inexpensively as possible. The operators' wages made up a large share of the company's expenses. Therefore the greater the productivity of the operators and the lower their wage, the more the company would profit. Promoting clubs and socials and instilling a family spirit helped to keep the union and collective bargaining out of the traffic department. The relatively small amount invested in paternalistic gestures proved to be profitable because it helped block the return of the union.

The union might have been kept out of the traffic department at least until the 1940s, had it not been for the outbreak of World War I and its effect on the economy and the labour movement. Until 1913 the organized labour movement in Canada had grown steadily. Its membership expanded from 50,000 unionists at the turn of the century to 175,000 in 1913. The first years of war marked a decline in the labour movement as many unionists enlisted. To make matters worse, the depression at the beginning of the war hit the highly-organized trades—such as printing and construction—the hardest. The trough of this depression for telephone workers was 1915, when the company reduced operators' worktime by one week a month, with a substantial reduction in pay. The craft workers saw their rates drop by 25 to 44 cents a day.

By 1916 the British Columbia economy began to benefit from wartime demand for metals and lumber. As well, the enlistment of workers shrank the labour pool, creating a shortage of labour by late 1916. In the summer of 1917 the telephone company suffered its first serious shortage of operators. Previously, during the 1902 and 1906 strikes, there had been temporary shortages, but the prestige of being a telephone operator and the limited job possibilities for women ensured sufficient new applicants. However, wartime labour shortages temporarily opened many jobs to women previously held exclusively by men.

These favourable market conditions saw a resurgence of the labour movement, with workers launching a new organizing offensive. With labour in high demand, the employers' open shop offensive was stalled. At BC Telephone the craft workers had successfully restored their 1913 rates by 1916. In addition they bargained for the 44 hour week, winning Saturday afternoon as a half holiday. The company, squeezed by the growing shortage of skilled workers, was willing to sign a contract with the men, the first since the ill-fated 1906 agreement. The agreement was for one year and was automatically renewed thereafter unless either party gave 90 days' notice in writing of its desire to terminate it. The company hoped that the agreement would continue in force throughout the war, but the union was dissatisfied with the 1916 agreement and sought a new one after a year.[32]

The workers were concerned with union security, the closed shop, and the company's use of workers in lower-paid classifications to avoid paying higher craft rates. In 1913 the company attempted to erode the position of craft workers by using apprentices. Although this move had been stopped by the successful 1913 strike, the company continued to unilaterally implement changes in craft classifications. The company tried to undermine the foremen, the highest paid craft workers who supervised a group of workers without actually using tools themselves. The company assigned "wire chiefs" to do the job of foremen as well as work with tools. Wire chiefs received a slightly higher rate than linemen, but not the foremen's rate. The company then created further divisions among the crafts by distinguishing between type "A" and type "B" wire chiefs, with different rates of pay. The 90 days' notice required to terminate the agreement was also contentious as the notice undermined the union's only real weapon, the strike.

After unsuccessful meetings, the company again turned to the IDIA and requested that a board of conciliation be appointed. The board met in the late summer of 1917 and an agreement was worked out and signed by fall. The union failed to win most of its central demands, including a limit on the use of wire chiefs and the closed shop. The new agreement, however, did include wage increases and a reduction in the notice period for termination of the agreement to 30 days. Both the company and the union recognized that the 1917 agreement was nothing more than a stopgap measure and that the issue of the closed shop was far from resolved. The union, strengthened by the fact that nearly all telephone craft workers were organized, decided on another drive among the operators. This marked a resumption of the battle for the hearts and minds of the "hello girls."[33]

At the beginning of the war the company held the advantage. The crushing of the operators' union in 1906 coupled with the company-oriented clubs, socials, and a general program of paternalism, made organizing difficult. In 1916 the company expanded its paternalistic program. It replaced the foundering company/employee-financed sick benefit fund with a totally company administered and funded employees' benefit plan. One year later a holiday camp for the operators, called "Oaks Point," was constructed at Buccaneer Bay. In November, 1917 operators received a small wage increase. A few months later the company created the new position of "welfare superintendent." The chief operator at Seymour was appointed to the position and employees were assured that "at any time anyone may take up personal subjects

Operators at the main exchange in Vancouver, turn of the century.

Vancouver City Archives

with [the welfare superintendent] privately, knowing that they will be treated as confidential."[34]

The company's new flurry of paternalism did not pay off. The employees' benefit plan was an improvement over the previous charity, but it was still a matter of company benevolence and not a worker's right. The holiday camp, though successful at first, fell into disuse as the operators preferred "to go three or four days to Seattle rather than go up to the camp where none of their men friends were allowed."[35]

By early 1918 the company, aware of the union drive, was desperate to keep the operators out of the union. In March, 1918 traffic department supervisors went to the length of presenting the following address to the company president, William Farrell.

> To Our Officials of the British Columbia Telephone Company
> Sirs
> We the Chief Operators, Supervisors, Teachers and Operators of the Traffic Dept. of the BC Telephone Co. Ltd., wish to express our gratitude and appreciation to the Officials of the Company for the many kindnesses and considerations we have received from time to time.
>
> We realize and appreciate the splendid conditions under which we work, knowing that so few companies would take so much trouble and go to so much expense in trying to make their employees' work a pleasure.
>
> We thank you heartily, First for the splendid summer camp at Oak's Point which provides such an ideal vacation. Then for the Tennis courts which give so much pleasure in the summer time. Then for the generous Benefit Fund allowance which has proved a blessing to so many of our co-workers and lastly for the new short hour schedule recently inaugurated, with its accompanying increase of salary.
>
> We are proud of our company and will be very pleased if the officials will accept this as a token of our loyalty and best thanks.
>
> From the Girls.
> Vancouver, March 27th, 1918.[36]

Less than three months after the presentation of this enlightened address, most operators, including many supervisors, were organized into IBEW Local 77A, The Telephone Operators' Union.[37]

As the telephone industry spread throughout North America, the IBEW began a major organizing campaign. The union added women organizers to the international staff. While the international was not yet ready to enlist operators directly into the craft locals, telephone work had become a significant area of employment and the union recognized the need to organize operators.

Vancouver City Archives

Laying conduit for underground cable.

Different structures for operator locals, sub-locals and auxiliaries were discussed and tried. Finally, in 1918, the international union decided to form a self-governing, financially-autonomous operators' department of the IBEW.[38]

Local 213 in Vancouver had previously been successful in organizing operators. It was not surprising that once they had re-established the union in the telephone company, they turned again to the plight of the operators. In June the craft workers invited the Vancouver operators to a special meeting of Local 213, addressed by a Seattle IBEW organizer, a woman named Odlin. More than 150 operators attended the meeting. A number of supervisors who were attempting to form a company union were also at the meeting and a debate broke out between Odlin and these supervisors on the relative merit of a company versus worker controlled union. By the end of the meeting the majority of operators had signed up with the workers' union.[39]

During the summer of 1918, Local 213 began an unprecedented attempt to negotiate a blanket closed shop agreement with the three city utilities — light, power and telephone. Local 213 had won recognition at BC Electric in 1907 and at the Western Canada

Power Company in 1912. But different rates, different conditions, and even different hours of work prevailed at these three organized worksites. The union wanted an across-the-board $6 rate for journeymen and a $7 rate for cable splicers, as well as the closed shop. All three utilities were presented with similar contracts and given strike notice. The telephone company settled a day before the strike deadline, signing an agreement which at last restored the closed shop. Brief strikes at the other utilities ended with the union winning its major demands.[40]

This was a sweet victory for Local 213. The employers' offensive against the closed shop had been broken. Through the blanket utilities agreement, the union had established city-wide rates for electrical work. The crowning achievement was that BC Telephone—a leader in the employers' offensive—had been the first to crack under the threat of a strike.

But the 1918 closed shop agreement with the telephone company once again contained a crucial weakness. This agreement repeated the dangerous error that had been instrumental in the union's devastating 1906 defeat: it allowed the company to separate the operators from the craft workers. The women had just been organized. The union felt that the operators would not be able to weather an immediate strike and decided that if a strike against the telephone company proved necessary, only the craft workers would be called out. Local 213, anxious to avoid a strike where some members would be out while others remained at work, accepted the company's offer of a closed shop for the craft workers only.

Chapter 3 *General Strike*

The world war which broke out in the summer of 1914 greatly affected the Canadian labour movement, particularly in the west. In the years preceding the war a rift had developed between eastern and western affiliates of the Trades and Labour Congress (TLC). The response of the federal government to the war widened these differences. By 1919 arguments over labour's response to the government's wartime policy led to an open split, which divided the labour movement for over 35 years.

The split had two important effects on telephone workers. It led indirectly to a separate, isolated telephone local of the IBEW in the twenties, which contributed to the rise of company unionism. Later this division in the labour movement presented a major barrier to the formation of a single union of Canadian telephone workers.

The labour movement in British Columbia tended to be strong, radical and socialist. The influence of conservative craft unions was weaker in British Columbia than in eastern Canada. The province had a large immigrant population steeped in the more radical labour traditions of Europe and Great Britain. The provincial economy was based on large-scale extractive industries like mining and forestry. These dominant industries—particularly susceptible to extreme market fluctuations—were the scene of major battles between employers and workers, often taking place in company towns and involving family and community in the fight. Employers were vehemently anti-union and they received the backing of the provincial government in their strikebreaking actions. Provincial governments in the "company province" saw their role as ensuring a good business climate. Government use of militia, the police or the army on the side of the employers was common. The intervention of the provincial government on behalf of employers convinced the British Columbia labour movement that the government was not a neutral force but clearly on the side of the employers. This political insight spurred the labour movement into political action and encouraged labour support of labour and socialist candidates. A wide spectrum of socialist ideas were current in the labour movement.

The problems of organizing a large industrial workforce in the face of hostility from employers and government convinced many unionists of the need for "industrial unionism" as opposed to "craft unionism." Most workers were unorganized. The small minority which was organized were in "craft" or "trade" unions, which organized only the skilled workers in a particular craft. These craft workers were able to win recognition and improvements in working conditions because they could not be replaced as easily as "unskilled" workers. But the vast majority of workers needed a form of union which could organize all workers in a particular industry or work area. This need gave rise to various attempts to establish industrial unions.

Western labour argued in favour of industrial unionism at conventions of the Canadian Trades and Labour Congress — the central labour body to which a majority of unions were affiliated in Canada, and the sister organization to the American Federation of Labor. Opponents of industrial unionism argued the futility of trying to organize the mass of workers. Craft workers, they argued, achieved their power through the skills of their members and through limiting admission to trades. Crafts such as the IBEW had been formed explicitly to "rescue the electrical trade from the conditions of an unskilled trade to which it had fallen." The leaders of the craft unions argued that their members were a cut above the unskilled. They had risen above the industrial worker and were not about to open their unions to the mass of unskilled workers. This would flood their organizations and eventually sink the privileged position of the skilled workers.[1]

The east-west, conservative-radical division in the labour movement extended through many of the international unions. In 1908 the IBEW split between advocates of industrial unionism and advocates of craft unionism. For six years there were two rival IBEWs, each of which claimed to represent the union. The factions were named after their respective presidents and were known as the Reid and McNulty factions.[2]

Locals 213 and 230 supported the radical industrial union faction led by J. J. Reid. The choice was an easy one. Local 213 had organized telephone workers on an industrial basis since the founding of the local. The craft workers recognized that the support of the non-craft operators was vital to an effective union in the telephone industry. As well, the main contact that the local had with the international was through other Pacific coast locals and the Pacific district office in San Francisco. These western locals were solid Reid supporters.

The split in the IBEW caused tremendous dissension in the American Federation of Labor. Various unions sided with one faction or the other. Both factions sought acceptance from the AFL, recognizing that its decision would keep one faction within the AFL and condemn the other to isolation. While there were some unions outside of the AFL, both factions were hesitant to go it alone. Electrical workers often needed the support of other crafts which would only be extended to the AFL-endorsed faction. A decade before, the AFL had adopted a policy of denying membership and support to dual unions—unions that organized workers within the jurisdiction of an existing affiliate to the AFL. Dual unions, it was argued, weakened the solidarity of the labour movement, opening the way to competition among unions for members. The concept of one craft—one union was fundamental to the craft federation. The group recognized by the AFL would retain the IBEW jurisdiction over all electrical work. The other organization would be expelled from the federation and excluded from city labour councils, which were chartered by the federation. In Canada the TLC, dominated by international unions, would undoubtedly side with the AFL decision.

In British Columbia the split had little immediate impact. When the AFL extended official recognition to the conservative McNulty group, the IBEW locals continued to be seated at meetings of the Vancouver and Victoria Trades and Labour Councils with the councils turning a blind eye to the dual unionism charge. The minutes of Local 213 show that communications were received from both factions. For many years the local simply threw the correspondence from the McNulty faction in the garbage and treated the Reid correspondence as the official communications of the IBEW.

The end of the factional struggle came in 1914, when the west coast locals of the IBEW voted to close ranks with the McNulty forces. But the split left a legacy of bad blood between the international office and the Pacific coast locals. For years after reunification the international office kept a wary eye on the Pacific coast for signs of another secessionist movement or any other symptom of resurgent "industrial unionism."

The outbreak of war sharpened conflicts within the labour movement. Under the influence of western delegates, the TLC had adopted a policy in opposition to the war and in support of a workers' general strike in the event of an outbreak of hostilities. But the congress quickly retreated from this position when war broke out, reneging on its opposition and refusing to organize or call a general strike. By-passing and conservative TLC leadership, many western locals conducted their own vote on the general strike. Telephone workers in Local 213 voted in favour of the general strike, but the idea was scrapped when it became apparent that without preparation and widespread support, the strike would be a complete failure which would leave the participants isolated and open to victimization.[3]

One of the most controversial government actions during the war was the introduction of compulsory military service. The conscription program began with a national registration of the workforce, followed a year later by conscription. Although it condemned conscription, the TLC went along with the registration program. In western Canada, however, the labour movement did not follow the lead of the TLC. Western labour organized against the registration program, recognizing it as a prelude to conscription. Western labour leaders felt that military conscription could lead to industrial conscription, with a consequent suspension of workers' rights.

Typical of many western locals, Local 213 opposed the mobilization. Local 213 passed the following motion at a membership meeting on January 3, 1917: "We oppose National Service registration until all honorary colonels are sent to the Front, and all profits from munitions are conscripted for the benefit of returned soldiers and their dependents, and the dependents of our fallen heroes." A common demand in the west was for the conscription of wealth instead of manpower.[4]

For those workers not conscripted, the government's wartime economic policies brought new problems. Inflation, price increases, and food shortages all started to appear in the late war years. The federal government outlawed hoarding and "artificial" price increases, but its refusal to monitor or enforce its own regulations rendered them ineffectual. Workers were asked to give concessions, sacrifices and accept continual speed-ups for the war effort while business rolled in increased profits. As inflation eroded working people's standard of living, there were an increasing number of strikes for wage increases. In the spring of 1917 there was a two-week strike in the shipyards and metal trades. Talk was that this strike might be expanded into a general strike, but it ended when the workers won a cost-of-living adjustment. On the heels of the shipyards strike came a strike of street railway workers, which shut down Vancouver's street car system for ten days. A year later the electricians won their unprecedented utilities agreement with the telephone, power and light companies.

In response to the rising labour militancy, the government adopted more restrictive policies and used unprecedented powers against labour. In the spring and summer of 1917 it passed a series of orders in council introducing conscription, widespread restrictions on freedom of speech, assembly, and the press, the banning of strikes and lockouts, and the extension of the Industrial Disputes Investigation Act into new areas. Ottawa even banned the newspaper of the BC Federation of Labour, the *Federationist*. Many radical organizations and their newspapers were banned. The government threatened industrial conscription in industries essential to the war effort, as many labour leaders had predicted. Local 213 responded by advocating a complete work stoppage. A motion was passed by the local in January, 1918 stating, "That this local go on record as being heartily in favour of a 'down tools' policy in the event of the government trying to put in force any legislation bringing about industrial conscription, or the introduction of indentured labour in Canada."[5]

By 1918 many locals in western Canada had canvassed the general strike, either at the outset of war or as a tactic to be used in response to government repression. Shortages of labour in the late war years had seen an unprecedented growth in labour organization which swelled the ranks of most unions. The Vancouver Trades and Labour Council was a centre of activity, coordinating organizing campaigns, solidarity actions, and opposition to conscription. This fervid activity gave cohesiveness, coordination and solidarity to the Vancouver labour movement.

In the summer of 1918 news of the shooting and death of former BC Federation of Labour vice-president Ginger Goodwin sparked the city's first general strike. Goodwin had been an organizer for the Mine, Mill and Smelter Union. Although he had been declared unfit for military service in 1917, Goodwin was reclassified as fit shortly after he led a strike in Trail. He decided to dodge the draft by going into hiding on Vancouver Island, where a special police constable shot and killed him. Goodwin's death shocked the labour movement and led to spontaneous walkouts in protest. Workers considered Goodwin a martyr to his trade union and anti-conscription beliefs.

The Vancouver Trades and Labour Council on August 2, 1918 declared a 24-hour memorial work stoppage for Ginger Goodwin.

The strike was successful, with the overwhelming majority of the city's labour movement staying off the job for the day. The strike was marred by one violent confrontation, triggered when some businessmen persuaded returned soldiers that the strikers were pro-German and anti-patriot. The veterans ransacked the Labour Temple, destroyed records, and assaulted officials of the Trades and Labour Council.

The telephone operators were exempted from the protest. The prevailing view in the local was that operators should not be asked to join in this militant action as they were far too inexperienced. The craft workers probably underestimated the operators' resolve. The only operator whose role in the strike was recorded was Francis Foxcroft, a member of the operators' Local 77A and telephone operator at the Labour Temple. When the mob of veterans ransacked the Labour Temple, they attempted to throw council secretary Vic Midgley out the window. Foxcroft, however, stood in front of the window and prevented the crowd from killing Midgley.[6]

In spite of the confrontation with the veterans, the strike had given a great boost to the city's labour movement. For one day workers had struck together in a solidarity action. But this protest action had not been a valid test of the general strike. The purpose of the strike had been to down tools for the day and demonstrate labour's anger over the death of Goodwin. There was no demand to the strike, nor any issue which could be negotiated. While some city labour leaders touted the one-day strike as proof of the power of a general strike, a few recognized the implications of the limited goals of the strike.

In the spring of 1919 the Canadian labour movement split. Frustrated by the conservative eastern leadership, western labour met separately in a Western Labour Conference held in Calgary, Alberta. Originally planned as a conference to prepare for the Trades and Labour Congress convention, the Western conference quickly took on a dynamic of its own. It passed resolutions on a variety of topics and condemned government censorship and repression. The conference resurrected the issue of industrial unionism and called for a referendum on the launching of an industrial union, the One Big Union (OBU).

The OBU was to be set up by reorganizing existing unions "along industrial lines, so that by virtue of their industrial strength, the workers may be better prepared to enforce any demand they consider essential to their maintenance and well-being." In place of conservative, protectionist craft unions, the OBU was to be an aggressive, all-inclusive, industrial union drawing together the unskilled as well as the skilled in a single organization. The industrial weapon of the OBU was to be the general strike.[7]

Almost immediately the Vancouver and Victoria Trades and Labour Councils endorsed the OBU. The Vancouver council voted a two-cent-per-capita assessment to pay for an OBU organizing drive. But before the OBU could be formally established—in fact before the ballots were returned on the OBU referendum — a general strike broke out on May 15, 1919, in Winnipeg.

The Winnipeg general strike stirred workers across the country. As word of the strike spread, messages of solidarity from unions and councils poured into the city. Labour newspapers sent correspondents to Winnipeg, and messages from the city's workers were read at Trades and Labour Council meetings in cities across Canada. Unions called rallies and information meetings to explain the strike and show solidarity with the Winnipeg workers.

The general strike alarmed the federal government. The success of the 1917 Russian revolution had left the government in fear of Bolshevism in western Canada. Radical unionism in the west and the growing sympathy for the Russian revolution among western labour leaders fuelled the government's fear. Many of the restrictive orders in council passed in 1917 and 1918 were specifically geared to deal with this perceived threat. The government viewed the Winnipeg strike as a usurpation of civic power through insurrection. The magnitude of the strike gave the strike committee the power to control services in the city. The demands of the strike and the actions of the strike committee, however, included little that went beyond simple collective bargaining. The goals were neither political nor insurrectionary.

The federal government did not wait long before intervening in the Winnipeg strike. After mounted police and militia units equipped with machine guns were dispatched to the city, the government issued an ultimatum to striking postal workers demanding that they return to work or lose their jobs. The postal workers voted to ignore the ultimatum and were fired. This action provoked a series of strikes in solidarity with the Winnipeg workers. On May 26, workers in Edmonton struck. Within days they were joined by strikers in cities and towns across the country. By May 30, 75,000 workers were on strike from Rossland, British Columbia, to Amherst, Nova Scotia.[8]

Vancouver workers were particularly alarmed by the federal government's intervention. The May 23 meeting of the Vancouver Trades and Labour Council gave the executive the authority to call a general strike if the military intervened in Winnipeg. But when word reached Vancouver that the Winnipeg postal workers had been fired, the executive called for a vote on a sympathetic general strike.

On June 1 the IBEW, in a special meeting, voted down the motion to participate in the general strike. The telephone craft workers felt that their services were essential and therefore tended to vote against strike action. But the local was divided on the issue. The issue remained controversial until, at a subsequent meeting, the results were reversed and the electricians went on strike, with the telephone craft workers bowing to the decision of the majority. The Vancouver sympathetic general strike started on June 3, 1919. Unlike the one-day protest action, workers had no idea how long this strike would last. There was also much confusion over which services were deemed essential and therefore to be maintained, and which services were to be shut down. The council had little time to prepare for the action with only informal communications and liaison between the council executive and the locals. As a result the Vancouver strike did not begin with solid support among all trade unionists. In fact the strike began to build up momentum

Cartoon in the June, 1919 *Strike Bulletin* attacking scabs. Vancouver operators were the last to return to work in the 1919 general strike.

UBC Special Collections

only after much discussion on different job sites and at union meetings finally persuaded the reluctant locals to join the solidarity action.[9]

In many ways the Vancouver strike was more radical than the precipitous strike in Winnipeg. While the Winnipeg strikers were supporting workers engaged in a struggle with the local captains of industry, the Vancouver strike was remarkable in that it was motivated by solidarity for workers more than a thousand miles away. Also, the demands of the Vancouver strikers were political; they were addressed to the government, not to employers. The demands went far beyond mere wage issues and dealt with wide-ranging social questions which included reinstatement of the Winnipeg postal workers and immediate settlement of their grievances, guarantees of the right to bargain collectively, pensions for soldiers and their dependents, compensation for veterans who had served overseas, nationalization of all food storage plants in order to eliminate hoarding, and the enactment of legislation mandating the six-hour day in all industries where unemployment was prevalent. The Vancouver workers, by entering the fray, were adding a new dimension to the Winnipeg action.[10]

A strike committee was set up with the business agent for Local 213, Teddy Morrison, as a member. The committee exempted some unions from strike action in order to avoid charges that the strike committee had usurped the power of the city government. Emergency services provided by firemen and policemen, as well as milk and hospital deliveries, were all exempted. The telephone network could not function without the operators, so telephones were included in the list of emergency services to be maintained and the operators were not called out.

Among the first to respond to the strike call were the shipyard and dock workers, who shut down the Vancouver waterfront on the first day of the strike. While most members of the Typographical Union favoured joining the protest, their constitution required a two-thirds majority before they could take strike action. As a compromise, they remained at work but set up a censorship committee "to ensure the publication of the strikers' views and [to prevent] deliberate misrepresentation . . . under penalty of cessation of work." By the second week of the strike the Typographical Union called a five-day suspension of the Vancouver *Daily Sun* as a reprisal for its anti-strike editorials. An issue of the Vancouver *Province* also failed to appear when the workers refused to print anti-strike advertisements.[11]

An important union which did not initially join the strike was the Street Railway Workers' Union. Its members, however, reversed their decision against participation within days and joined the walkout on June 5. This walkout brought the first counter-measure of the strike by city officials. City council, in its attempt to undermine the shutdown, rescinded the anti-jitney by-law, which banned the use of passenger-carrying motor vehicles.[12]

This action began an escalation of the struggle. The strike committee called out the unionized civic employees. This spread of strike action prompted a city council ultimatum ordering all civic employees to return to work or be fired. The strike committee remained firm and threatened further strike action. If council refused to reinstate the anti-jitney by-law or if it took disciplinary action against the civic employees, the telephone operators would be called out on strike.

Council ignored the strike committee's announcement, and on the morning of June 14, 325 telephone operators and their immediate supervisors joined the strike. On the first morning of the strike at Seymour all that remained of the Vancouver toll exchange staff was one chief operator, one supervisor, two toll operators and the traffic engineer. As the women left telephone offices in outlying areas, they locked the doors and dropped the keys through the window as they left.[13]

The telephone operators' walkout marked the peak of the strike. It was the last offensive action by the strike committee. With most of the city's organized workers off the job, the strike committee had played its highest card. The workers' strategy had been based on the belief that a mass walkout by organized labour would bring about a quick victory. This level of confrontation would be difficult to maintain without immediate results. When council, bolstered by the business community, refused to budge, the strike committee had no further initiative. From this point on, the committee was forced into retreat. It dropped its demands on the federal government and switched all efforts to the defence of the workers who had gone on strike. The goal was to find a method for returning them to work and avoid victimizations.

With the first walkout, anti-strike forces also started to organize. A week later a group of "concerned citizens," dedicated to maintaining "law and order," formed the Vancouver Citizens' League. With support from the business community and employers' organizations, the Citizens' League published an anti-strike paper which condemned the strike for being an insurrection and a conspiracy stirred by socialists and communists. The Citizens' League bought space in the Vancouver dailies to place advertisements aimed at discrediting the strike. The league recruited strikebreakers for the jitney service and later for the telephone company.

The telephone company claimed that with the aid of the

Citizens' League, 76 strikebreakers were eventually hired at the telephone offices. Most were society women doing volunteer strikebreaking and ex-operators attracted by the generous pay the company offered for strikebreaking.[14]

As the strike extended from days into weeks, the telephone operators' difficulties became acute. The meagre financial resources of the local were soon depleted. The operators' local had no strike fund, so most operators had to work out temporary arrangements to pay for rent and board. The city's labour council set up a fund for the operators, and wealthier unions sent donations. This money was distributed to the more needy operators, but there was not enough to go around.[15]

Leona Copeland, who started as an operator at BC Telephone a year before the strike, recalls the financial strain, noting that it was particularly hard for single operators:

> My landlady kept me going. I wasn't right down to brass tacks, but pretty close to it. She kept me going by not coming after me all the time, looking for her rent money. Most of us who stayed out, couldn't afford to stay out, but we did. I don't know how we did it.[16]

By the third week the strike began to deteriorate as fear of victimization spread. On June 19 the CPR workers voted to return to work. The following day about a third of the striking civic employees abandoned the strike. Soon groups of woodworkers, teamsters and brewery workers went back, and a few mills and breweries started up again. Just when it appeared that the strike would dwindle away, word reached Vancouver of the arrest of the Winnipeg strike leaders and the violent confrontation in Winnipeg between the mounted police and strike supporters on "Bloody Saturday," June 21. These events rekindled support for the Vancouver strike, and Victoria joined the sympathetic strike on June 23. But the renewal of momentum was short-lived. Winnipeg ended its general strike on June 26; Victoria went back a day later.

The Vancouver strike continued for a week after the crushing of the Winnipeg strike. Vancouver strikers insisted that there be no victimization as a condition for the return to work. The demand was aimed primarily at the city government. A third of the city's striking employees had abandoned the strike and successfully returned to their jobs. The strike committee wanted formal assurances that the rest could return. But as the days wore on, groups of workers negotiated their own back-to-work agreements, by-passing the strike committee. On June 30 the remaining civic employees out on strike returned to work.

By the beginning of July the strike committee decided to end the disintegrating strike. There remained one group of strikers still having difficulty in negotiating a no-victimization return to work—the telephone operators. The telephone company took the position that while operators would be rehired, striking supervisors would be demoted to ensure that only anti-union strikebreakers held supervisory positions. The women felt that they could not return under such terms. They decided to continue their strike in order to defeat the company's attempts to victimize the striking supervisors.

The telephone craft workers agreed to stand with the operators, and both decided to release the rest of the labour movement from any obligation to continue the strike. On July 3, 1919—one month after it had started—the Vancouver general strike was at an end. The strike was the longest of the sympathy strikes, and the last to end.

In announcing the end of the strike, the labour newspaper, the *Federationist*, praised the resolve of the telephone operators. The article also commended the electricians and operators for releasing the strike committee from its commitment of no victimization:

> The action of the telephone girls in responding to the call for a general strike has placed them in a class by themselves amongst women workers in this province. With only a few backsliders, these girls have won the admiration of all those who admire grit and working class solidarity. That their action will be remembered by the workers not only of this city, but by the workers all over the continent for their loyalty, goes without saying. If all the men had displayed the same spirit, the strike could not have been finished with them carrying on their fight against discrimination, after the general strike was called off. The strike was called off as a result of the telephone girls and electricians taking the stand that they could fight the matter of discrimination against the telephone operators alone. This decision was arrived at on Wednesday night after a three and a half hour meeting, at which a resolution was passed urging the other organizations to return to work, as they were of the opinion that Local 213 of the Electrical Workers and the Telephone Operators could fight the discrimination matter out themselves. This was announced at the mass meeting held in the Labour Temple on Thursday, and as a result the strike was called off at 5 pm.[17]

The operators continued their fight with the company for almost two weeks. By mid-July the union was forced to accept what were essentially the company's terms. While the company agreed to take back all employees without discrimination, striking senior operators and supervisors had their status reduced to that of operator. A few of the supervisors were to be reinstated in their old jobs within 30 days. The remaining ones were to be first in line for

An early BC Tel repair shop. *Vancouver City Archives*

vacant supervisory positions. The men returned to work on July 15 and the women returned a day later.[18]

In the months following the general strike the labour movement was in turmoil. The One Big Union had been launched in the midst of the strike. The new industrial union was popular in the west. Its influence spread rapidly, and by the end of its first year the OBU had 29 locals in British Columbia and a membership of close to 20,000.

The OBU was opposed by government, the employers and the international unions. Both government and employers saw the new union as a far more serious threat to industry than the craft unions. The OBU was closely identified with the general strike and many leaders of the OBU were socialists. The internationals considered the new organization to be a "dual union" and they were not about to let it raid their membership. Organizers from international unions were sent west to stop the growth of the OBU.[19]

The Canadian vice-president of the IBEW, Ernest Ingles, was furious when he learned that Local 213 had participated in the Vancouver sympathetic strike. The workers in Vancouver had one of the few closed shops in the telephone industry. By joining the sympathetic strike, the local not only jeopardized its closed shop, but had also endangered the operators' local. The international, following craft traditions, was interested in distinguishing between telephone workers, wiremen and linemen. Local 213's traditions, however, had more in common with industrial unionism than with narrow craft allegiance. Though Local 213 had been chartered as a lineman's local, by 1919 it had a membership of 600, which included a variety of electrical workers. The international feared that Local 213 was again on the road to secession and industrial unionism. Ingles sent a message to the local on June 22 ordering it to quit the general strike and return to work. The local ignored Ingles' order, and it won the dubious distinction, in the eyes of the international, of being the last union in Canada to return to work after the "OBU strike of 1919."[20]

The Vancouver Trades and Labour Council split shortly after the 1919 strike, with the international unions withdrawing from the OBU-dominated council and establishing a council of their own. Local 213 was ordered to quit the Vancouver Trades and Labour Council (OBU) and join the new council. Local 213 refused at first, though eventually it moved with the majority of international unions into the new council. These two acts of defiance by the local landed it in hot water with the international. Ingles, on behalf of the international, suspended the local, which decided to appeal to the international. It passed a motion to "carry on business of Local 213 as though no International Officer had given any decision against us."[21]

The international convention held in the early fall turned down Local 213's appeal and issued a new charter, Local 310, for Vancouver electrical workers. The international ordered the members of the suspended local to join the new local, but again the majority of electrical workers refused. They continued to pay dues, hold meetings and conduct business in the name of Local 213.

The telephone craft workers, however, saw the creation of Local 310 as an opportunity to establish an independent telephone local. The craft workers felt that Local 213, which included workers employed by a variety of companies, did not devote enough attention to the problems of telephone workers. They were also angry at the electricians at BC Electric Company, who had urged the telephone workers to join the general strike but had returned to work, leaving the telephone workers on their own.

Most telephone craft workers were by now permanent employees of the telephone company. They no longer needed the local to assist them in finding other jobs in the electrical field. Most jumped at the opportunity to leave Local 213 for the new local. Local 213 saw this as a betrayal, which aggravated the already present antagonism between the telephone workers and other electrical workers.

When Local 213's appeal to the international failed, the local took the issue to court. Trade unionists are usually reluctant to bring internal disputes before the courts, but the membership of Local 213 felt that the actions of the international indicated that it was "the members, not the charter" which were under suspension. Local 213 also wanted to establish its right to the resources of the local, which amounted to nearly $14,000. In their view the courts were the only avenue of appeal left open to them.

The court case was the first of its kind and was decided in favour of the local. The judge ruled that the charter of Local 213 had been improperly suspended and revoked. He ordered the international to restore full IBEW membership rights to the local. The judge concluded that the international had no claim on the property of the local. An injunction was served, restraining vice-president Ingles from interfering with Local 213's charter.[22]

The international decided not to appeal the court's decision. By this point, the spring of 1920, it was clear that Local 213 was not heading into the OBU, so the international decided to try and patch things up with the local. Ingles tried to get the locals to amalgamate, but these attempts failed because the telephone workers wanted their own local. Furthermore, Local 213 wanted to discipline the members who had joined Local 310. The resulting jurisdictional crisis affected the union's existing contracts. At both BC Electric and BC Telephone Local 213 had established a closed shop, but when Local 310 was created, the BC Telephone agreement was handed over to the new local. BC Telephone declared that the closed shop agreement meant that it could employ only members of Local 310 and it laid off members of Local 213.[23]

Recriminations flew back and forth; both locals sent appeals and demands to the international asking that the other local be suspended or its charter revoked. By the summer of 1921 it had become obvious to all that the two units could not coexist within a single local. The international decided to allow Local 310 to represent the telephone workers, while Local 213 retained the rest of its previous jurisdiction.[24]

The operators' local was totally demoralized after the general strike. The strike had taken a greater financial toll on the operators' local than on any other organization which had participated in the strike. The operators were among the lowest paid workers in the organized work force and they had been on strike longer than any other group of workers in the city. Probably more demoralizing than the financial problems, though, was the internal strife within the labour movement after the general strike. The spirit of 1918 and 1919—when the operators organized and felt part of a surge of solidarity among working people—had vanished. The Winnipeg strike had been crushed and its strike leaders jailed. The Vancouver labour movement was split between the forces supporting international unions and those backing the OBU. The IBEW itself had split, with the telephone craft workers battling Local 213, Local 213 battling the international, and the operators left to fend for themselves. At the exchanges operators were being supervised by strikebreakers who owed their position to their having worked during the strike. The unity of purpose and feeling of strength that had been so evident before and during the general strike had disappeared from the labour movement.

In the months following the strike there were a number of attempts to strengthen the operators' union. But these were limited and paternalistic. A lounge for the operators was opened at the union hall on Pender Street, just a few blocks from the Seymour exchange. Taking a leaf out of the company's book, the union furnished the lounge with "easy chairs, couch, piano, writing tables, nice curtains at the windows and a vase of flowers on the centre table." A "Jolly Girls Club" was set up to hold dances every second Friday. The union hoped that activities centering around the new lounge would encourage those operators who "were

The operators' lounge.

Vancouver City Archives

staying out of the union because of company influences" to join the union. These attempts failed. The labour movement as a whole went into decline in the twenties, and the operators fell victim to this disintegration. Late in the twenties the local formally folded. The operators were once more without a union.[25]

The craft workers in Locals 310 and 230 (Victoria) were more fortunate; they signed a closed shop agreement with BC Telephone. In this period of rapid expansion of telephone service, the company was anxious to hang on to its experienced craft workers. BC Telephone recognized that in the short term this meant accepting a union. The company tactically chose to support the separate telephone local, recognizing that an isolated local—with members in only one industry—would be easier to manipulate than the radical industrial OBU or Local 213 with its construction and power workers.

From a craft perspective, developments in the industry tended to support the move to a separate local. Telephone work had become increasingly specialized in the decade preceding the war. Standardized procedures and equipment had created a new telephone craft which required skills distinct from those of the construction electrician or the power electrician. The early telephones had been bulky, undependable instruments which needed large individual batteries to operate. By the end of the war they had become sleek and reliable, and ran on a common battery from the telephone office. More significant than these modifications was the change in popular attitude towards telephones. In 1902, when telephone workers first organized, there were no more than 2,000 telephones in the province. These were installed in business offices and in the homes of businessmen and were primarily for business use. The average person rarely used or came into contact with a telephone. By the twenties, however, there were more than 60,000 telephones in use in the province, and the telephone was no longer a luxury of the rich but a general household fixture.[26]

The twenties saw unprecedented growth in the industry. By the end of the decade there were 114,000 telephones in British Columbia, and by 1925 the province led the country in telephone use with 18.1 telephones for every hundred inhabitants. This surpassed Ontario's rate of 17.4 and was well ahead of any of the other western provinces.[27]

Paralleling the growth in the number of telephones, the long distance connecting radius of the telephone also increased greatly. Long distance calls were still remarkable events, and the Montreal to Vancouver connection in February 1916 was a major event. It was staged at the Globe Theatre in Vancouver and the Ritz-Carlton hotel in Montreal. Newspaper reporters and the public in both cities crowded into the respective centres to witness the feat of speaking to someone 3,000 miles away. Most of Canada's long distance calls, like the Montreal link-up, were routed through the United States. Vancouver was not linked to Ottawa by telephone until 1920, and even then the call was routed through Seattle, San Francisco, Salt Lake City, Denver, Chicago and New York. At each centre an operator had to route the call to the next exchange. Such a call involved many people and often took hours to set up. In 1921 the telephone companies across the country banded together to form the Telephone Association of Canada. The association formed the alliance necessary for an "all-Canada" telecommunications link. However, it took a further eight years until the project was finally adopted.[28]

By 1926 BC Telephone had a permanent telephone work force of more than 2,000, the vast majority being telephone operators. As the number of operators grew, the company developed several distinct operator classifications based on the traffic that each exchange handled, with different wage scales in "A," "B" and "C"

The auditing department. The IBEW, the craft union, did not organize the small clerical staff.

Vancouver City Archives

exchanges. A separate division was set up to handle the fast-growing toll service. Toll calls usually required more operator time than local calls. The operator also had to file a "ticket" recording details of the call for billing purposes. The tremendous volume of tickets at major exchanges led to the creation of a special clerical staff which dealt solely with billing.

With the dramatic growth in the number of telephones and the parallel increased demand for service, the clerical work force grew rapidly. In 1904 two women clerical workers formed the core of the commercial office. They kept the accounts of all Vancouver subscribers, making out service bills, taking cash payments, making up the directory, taking orders for new telephones, and serving those customers who wanted to make long distance calls from the company office. By the late twenties these functions had to be delegated to special departments, each of which required a staff of a dozen people or more. Staffing the wickets at the commercial office in Vancouver, for example, required a dozen clerks during peak periods.[29]

Sales also became a separate department. For most residential customers the leasing of a telephone was straightforward, but for large businesses and offices a private branch exchange (PBX) was often required. PBX sales required a specially-trained sales staff familiar with the company's switching equipment and trained to deal with the needs of large businesses. By the mid-twenties there were 200 PBXs in the province, with the majority at railway stations, government offices, department stores and hotels.[30]

Craft work also became more diversified. Historically, work within the electrical industry was divided into three specialized skills: cable splicer, lineman and wireman. At the top of the craft was the cable splicer. Connecting cables of hundreds of wires, each requiring identification, demanded great skill. Closing a splice and assuring that no moisture was left in the line also required exacting skill.

Linemen were journeymen electricians who did the basic construction and maintenance of the outside plant—they strung the miles of wire connecting the telephone network. The wireman was the journeyman responsible for indoor wiring. In the early contracts reference was often made to apprentices, who worked with the journeymen to learn this trade. A final category outside of the craft but often included in the contracts were groundmen or labourers. These workers assisted craft workers in placing poles or doing the physical work that went with pulling wire into place.

In telephone work the craft divisions followed the general trade subdivisions. The telephone cable splicer earned the highest wage. Linemen received the rate for journeymen, which was the same for

BC Tel linemen. The craftsmen enjoyed good wages and conditions in the 1920s but neglected their union.

Vancouver City Archives

all electric utility craft workers by 1918. In addition to linemen the telephone company had two special categories which were paid the craft, or lineman's wage. These were installers and central office maintenance workers. These two categories were the telephone industry's specialized wiremen. Installers did the actual installation of the telephone at the subscriber's home or business. The central office maintenance men did the servicing, repair and maintenance work required to keep the company's switchboards and central office switching equipment functioning.

As telephone work became more specialized, the company increased its pressure to train its craft work force. While any electrician would have a general understanding of the theory involved in telephone work, a "telephone" craft worker familiar with BC Telephone and its methods was far more valuable.

The 1922 closed shop agreement between BC Telephone and IBEW Locals 310 and 230 suited both parties. The telephone men had their own local and the company a supply of skilled craft workers. But a telephone local separated the telephone craft workers from other electrical workers, allowing the telephone workers to fall under the influence of the company. By the late twenties most telephone craft workers identified more with the company than with the union; the closed shop existed only on paper. Few workers were interested in policing their collective agreement.

Things were good for telephone craft workers in the twenties. They enjoyed good wages throughout the decade and had few grievances with management. Those grievances that did occur were usually settled informally as the company encouraged employees to by-pass the union. Work was steady and the employees viewed their jobs as permanent; they felt they had attained prestigious jobs which would last a lifetime.

By May, 1926, the union decided to negotiate a new agreement. Wages had remained the same for four years and many workers felt it was time for an increase. BC Telephone informed the union that no negotiations could take place until the existing agreement signed in 1922 was discontinued. This agreement, initially for one year, was automatically renewed until terminated by either party on 30 days' notice. On the company's recommendation the union gave the required notice to terminate the agreement. The union then attempted to set up a meeting to negotiate the new agreement, but the company, to the astonishment of the workers, refused.[31]

The company had never intended to permanently accept the union and was again trying to crush it. But this time they had not used the crude force of strikebreakers or a lockout. With the carrot replacing the stick, the company had enticed workers into complacency and inactivity. In contrast to the confrontation and turmoil of the late war years and early post-war period, the mid-twenties were tranquil, with employers taking advantage of the demoralization of labour and rolling back the gains of a few years previous. With the OBU long since dwindled in size, the threat of industrial unionism had passed. As well, the telephone craft workers were now isolated in their own local. The company had enticed the craft workers away from their union; most members no longer participated in the union.

In June the company issued its "Synopsis of Working Conditions and Wages" and handed a copy to each employee. The new terms included merit pay. This scheme allowed the company to pay different rates to workers doing the same work according to their "efficiency and faithfulness." The union, shocked at the about-face in the company's attitude, rejected the company's scheme as an attempt to undermine their members' solidarity. The union elected a negotiating committee to meet with the company and held a strike vote to back the demands of the committee. But the company had chosen the right time to break the union; the union had become weak and demoralized. Within a month the strike vote was overturned. Unable to mobilize against the company's proposals, the union withdrew its strike threat and the negotiating committee dissolved. Merit pay was imposed on a divided, though not entirely unwilling, work force.

A few months after this defeat the maintenance workers tried to discuss holidays and shifts with the company. These negotiations also got nowhere. By the spring of 1927 enough workers had become concerned about the defeat of the union to turn, as a last resort, to the Industrial Disputes Investigation Act and seek the appointment of an investigation board. The workers felt that, with a closed shop at the BC Telephone for a number of years, the union's right to representation had been established. They believed that the IDIA board would use this precedent to order the company to commence bargaining with the union.

The IDIA board, however, sided with the company. It found that BC Telephone was not obliged to negotiate with the union, and advised the workers to build up their morale and accept the company's schedules until they had enough support to alter them. The investigation documented the craft workers' demoralization. Because their battles had apparently been won, they had abandoned the collective strength of their union. For a while BC Telephone workers had been the highest paid telephone workers on the continent; they had few grievances or complaints. Few realized that their relatively good position was a result of the battles they had won as a union. When the telephone workers gave up their union in the late twenties, few could foresee the long and hard struggle it would take to organize again.

Most workers could not understand why the company refused to allow things to continue as before. But the company's long-term goal, to free itself of the union, was unchanged. In 1926 the company got its opportunity. Testifying before the IDIA board, a company spokesperson stated that for the company it was "thought not in the best interest to have a contract" and that "there is no organization of telephone workers in the Pacific States." The company openly admitted in the hearings that they had never abandoned their goal of an open shop.[32]

After the loss of the closed shop and bargaining rights and the subsequent IDIA investigation, the telephone workers at BC Telephone dissolved the local. Some individual craft workers kept their IBEW memberships to maintain their union death and sick benefits; others did so in the hope that a new local would re-establish itself. But Local 310 folded in 1929, with the craft workers who still felt loyal to the IBEW transferring to Local 213 or Local 230. In the wake of the defeat of the IBEW a new organization emerged. This was the British Columbia Telephone Electric Employees Organization, a company union.

Chapter 4 *Company Union*

By the mid-twenties the BC Telephone Company was seriously short of capital. The popularity and improved technology of the telephone increased the demand for wider connection. The prospect of large toll revenues from cross-country and international services spurred BC Telephone to further consolidate its position as the major telephone company in the province. To eliminate competition in this highly profitable area, BC Telephone decided to take over the government-owned lines, which connected the province's community telephone companies, as well as some of the small companies themselves. But takeovers and expansion were expensive. Once again BC Telephone was searching for a buyer.

The Bell Company—the giant monopoly in eastern Canada—was approached first. The choice was logical since BC Telephone purchased the bulk of its equipment from Bell. Bell made an offer, but it was below the trading value of BC Telephone shares and was rejected.

The company next made overtures to an American utilities company headed by Theodore Gary. The offer was more to BC Telephone's liking and a deal was struck. Theodore Gary and Company was already one of the largest "independents." BC Telephone Company was to remain in the hands of Gary interests until the merger between Gary and General Telephone Company in 1955. The sale of the company to United States interests had little effect on the telephone workers. In fact most telephone workers in later years believed that the United States takeover of BC Telephone occurred in the sixties with the GTE changeover. The popular image of BC Telephone until the late fifties was that of a "family company" run by the Farrells.[1]

This popular image had some basis. A year after the sale of BC Telephone, William Farrell's son Gordon became president of the company. The Farrell family invested in the United States parent company, and they were given a relatively free hand to manage the local company. BC Telephone's image of a family company was extremely useful, particularly as it complemented the company's paternalist strategy for countering the union. The appearance of local ownership also undercut demands for nationalization and public control of the company.[2]

The Gary purchase nevertheless did bring about important changes. BC Telephone—now part of a large conglomerate—began to buy its equipment from Automatic Electric, the Gary manufacturing company. Automatic Electric produced radically different switching equipment which made possible the automatic exchange and dial telephones. This type of switching equipment, known as "step-by-step," eliminated the operator from the local call.

In the late twenties BC Telephone began converting a few communities—Hammond, Chilliwack, Victoria and Powell River—to the dial system. But conversions slowed almost to a stop during the depression and were not resumed on a large scale until well after the Second World War.

Shortly after the takeover, Gary purchased a number of the small community telephone companies, including Kootenay Telephone, Mission Telephone, and major sections of the Dominion Government system. These smaller companies were not all brought directly into the BC Telephone fold (most merged into BC Telephone in the fifties), but as subsidiaries of the same parent they were removed as potential competitors of BC Telephone. The purchases effectively tightened BC Telephone's control to a monopoly within the province, and the purchase of the long distance connecting lines of the Dominion Government system gave it control of the inter-provincial connection through Alberta.

The takeover had cost the Gary corporation little in hard cash. The purchase was financed by a share issue totalling $4.5 million; the company was purchased and the costs recovered by issuing new shares. The sale had provided Gary's subsidiary, Automatic Electric, with a captive customer and a profitable monopoly in British Columbia.[3]

By this time the company's terms and jurisdiction no longer fell under provincial legislation. The company was now federally regulated and empowered with specific rights under federal legislation. The change from provincial to federal jurisdiction was largely a response to political demands for nationalization and local regulation, accountability and control of the telephone industry.

The regulation, ownership and powers of telephone companies were major issues in Canada almost from the inception of the industry. Until 1905 the evolution of the industry in British Columbia was not distinctly different from development in other provinces. After 1905, however, BC Telephone, the giant Bell Company in eastern Canada, and companies in the prairie provinces moved in different directions. Bell too was brought under federal regulation by the Board of Railway Commissioners, and the three prairie provinces bought out the local Bell interests and established provincially-owned telephone companies.

Bell's arrogant disregard of local authorities and the concerns of municipalities and towns fueled demands for provincial ownership. As early as 1880 Bell had obtained permission from the federal government to construct and maintain telephone facilities throughout Canada. Bell's charter, granted by parliament, declared the work of the company to be "for the general good of Canada" and gave the company a "right of way" to construct lines through any province or municipality in Canada. Like the railway, Bell soon became the focus of popular dissatisfaction.

The telephone held out the promise of conquering the isolation created by the long distances and sparse population which prevailed throughout Canada, especially in the prairies and rural areas. But westerners found Bell reluctant to provide a telephone network in rural areas. The lucrative telephone market, of course, existed in the large cities and towns, where a few miles of wire could connect thousands of subscribers. It was there that Bell concentrated its efforts. Its goal was to capture the city networks and the highly profitable long distance toll lines between cities. Telephone service was quickly extended to most large towns. In the rural areas, where hundreds of miles of wire were needed simply to connect a few subscribers, Bell ignored requests for telephone service. The Bell Telephone Company was in business to make profits. As long as the rural service was unprofitable, Bell was not interested in providing service, the "general good of Canada" notwithstanding.

To placate rural demand for telephone service, Bell issued manuals and pamphlets explaining "How To Build Rural Telephone Lines." The idea was to encourage farmers to purchase equipment, build their own farm lines, and then connect with the wider Bell network. Bell would gain the long distance revenues without being saddled with the large capital expenditures necessary to construct and service rural lines.[4]

The prairie provinces decided to use revenues from the lucrative urban areas to subsidize the unprofitable but equally necessary rural lines. In 1908 the Saskatchewan government took control of the profitable long distance lines in order to establish and subsidize rural telephone cooperatives. By 1920 Saskatchewan had a telephone system that few provinces could rival. The province had 1,544 rural telephone cooperatives and the highest saturation of telephones in Canada.[5]

In 1905 opposition to Bell forced the federal government to form a committee, headed by the postmaster-general, William Mulock, to study the problems of the telephone industry. The Mulock committee uncovered a variety of unsavory tactics used by Bell to prevent competition, including predatory pricing, exclusive contracts with municipalities and railways, and denial of interconnection rights to competing companies. But before the committee could deliver its evidence and offer remedial recommendations, Mulock was appointed judge and replaced by A. B. Aylesworth, who had served as Bell's lawyer during the committee hearings. Naturally, after Mulock was replaced, the committee accomplished little. But one important action was taken: Bell was placed under the jurisdiction of the Board of Railway Commissioners so as to establish some form of public scrutiny over the company.[6]

Like Bell, BC Telephone came under attack for not providing rural service. As in the prairie provinces, rural and farm communities started their own telephone companies. BC Telephone bought many of the larger local companies during its first period of consolidation at the turn of the century. Following this initial consolidation came a period of rapid growth in the industry which left a proliferation of small companies in British Columbia.[7]

The campaign for nationalization and local regulation and control of the telephone company was as strong in British Columbia as in other western provinces. For years the city of Vancouver fought to gain the right to regulate the telephone company and assert some influence over its power within the city. Vancouver city council and the telephone company had a running war for many years over the company's digging up of city streets, the location of poles, and the clusters of overhead wire. But appeals to the legislature to amend the city's charter and appeals to the courts to rule against arbitrary acts by the company were unsuccessful. Vancouver was not the only municipal authority that fought the telephone company. In the interior, company employees were thrown into jail on charges of "willful damage" when they tried to dig up city streets to place telephone poles. But in all cases the courts ruled that the power conferred on the company by the legislature took precedence over the rights of the municipalities.[8]

From 1903 to 1915 the province was ruled by the Conservatives led by Richard McBride. McBride greatly assisted BC Telephone in its attempts to bar other companies from the province. For years the provincial government also balked at legislation to permit municipalities their own telephone companies. By 1912 the municipalities had changed tactics in favour of regulation of both the telephone and the electric street railways through a Public Utilities Commission.[9]

At first both the Western Power Company, which owned and operated the street railway, and BC Telephone successfully lobbied the provincial government to drag its feet on the Public Utilities Commission. But in 1915 the Western Power Company, under competition from jitney service, switched their stance on regulation. The company felt that regulation would help eliminate competition from the jitneys. BC Telephone, however, saw no potential advantage in provincial regulation and sought to avoid the regulatory commission altogether. William Farrell and other major shareholders petitioned the federal government to enact legislation incorporating a new telephone company, the Western Canada Telephone Company. They hoped to transfer their assets to the new company, and thereby by-pass provincial regulation.[10]

The company of course did not argue that its move for a federal charter was based on the desire to avoid provincial regulation. The company explained that it intended eventually to connect beyond the borders of British Columbia, which required a federal charter. In fact a federal charter was not necessary to link telephone lines across provincial or even international borders. Even today the prairie provinces are linked nationally and internationally without a federal charter. Also, BC Telephone already had a federal charter, through its subsidiary, Canada Western Telephone and Telegraph Company, which linked the southern interior to the United States.

The municipalities did not end their campaign for government regulation of BC Telephone; they simply shifted the focus of their campaign from the provincial to the federal government. They lobbied the federal government to ensure that BC Telephone, like Bell, would be regulated by the Board of Railway Commissioners. Their efforts were successful. In April, 1916, the Western Canada Telephone Company Act was passed carrying a clause that placed the company under federal regulation.

It is not clear why the company chose to opt for a completely new company charter rather than transfering its assets to its existing federally-chartered subsidiary. Possibly they feared this action would be too obvious. BC Telephone did not immediately transfer its assets to the new federal company, but by 1918 the Western Canada Telephone Company was leasing the company's assets within the province. A year later the name of the Western Canada Company was changed to BC Telephone Company. For a few years a confusing situation existed. The BC Telephone Company, which was federally chartered, owned the assets of the BC Telephone Company Ltd., which was provincially chartered. By 1923, however, the federal and provincial companies merged, bringing the original BC Telephone Company under federal jurisdiction. Under the threat of provincial regulation, the company had finally accepted that regulation of some form was inevitable. BC Telephone preferred federal jurisdiction because a federal regulatory body was not as susceptible to pressure from municipalities or the local population.[11]

Shortly after coming under federal regulation, the company applied to the Board of Railway Commissioners for its first rate increase. Previously the company's rate ceilings had been written into its incorporating legislation. In 1921 the commissioners awarded the company a 12 per cent increase, setting rates which lasted until 1950. Some municipalities sent representatives to the hearings to oppose the rate increase, but the nature of the hearings, which relied on the company's own evidence of its expenses and financial requirements, precluded a meaningful investigation into real costs.

The hearings revealed the fundamental weakness of a regulatory system built around experts drawn from the very industry being regulated. The expert witnesses called before the commission were management personnel from other telephone companies. Bell witnesses, for example, testified that BC Telephone's claimed depreciation rate was legitimate. Clearly a high depreciation rate established at the BC Telephone hearings would assist Bell when it approached the Board of Railway Commissioners. The composition of the commission itself was heavily weighted to the company's perspective. As company president Gordon Farrell noted:

> Mr. Carvell was a shareholder of the New Brunswick Telephone Company and he was Chief Commissioner of the Board of Railway Commissioners. Mr. Robinson was the President of New Brunswick Tel and was a friend of Mr. Carvell. We wooed Mr. Robinson, who talked to Carvell.[12]

By 1930 many of the features of today's BC Telephone already existed. The company was federally regulated and foreign owned. The company was by far the largest telephone company in the province, with little fear of competition. By the end of 1925 BC Telephone owned over 92 per cent of the telephones in use in the province.

The company had also destroyed the union, replacing the workers' organization with a subservient company union. When the economic depression began in 1929, BC Telephone was in a sound position to weather the bad times. The employees, on the other hand, were completely disarmed.

The British Columbia Telephone Company Electric Employees' Organization (EEO) had been established with the assistance of management in the mid-twenties by those employees willing to aid the company's effort to break the union. Art Marsden, a telephone installer who was in the IBEW at the time, referred to the EEO in its early years as a "secret society." According to Marsden, the EEO was strongest among the inside workers and generally opposed by the outside workers. Roy Thompson, an installer with BC Telephone at the time, referred to the early leaders of the EEO as "company yes men." They wanted to be a part of management and "get ahead." They had abandoned the union path of improving the conditions of all its members. The EEO offered aspiring supervisors an opportunity to get noticed by management and prove their leadership skills. Few leaders of the EEO did not eventually graduate into the ranks of management.[13]

In certain respects the succumbing to company unionism by the inside maintenance and central office craft workers reflected their privileged position. Unlike the outside workers, they were in daily contact with management. The central office men were increasingly tied to the telephone industry because their specific telephone skills were not transferable to other areas of the electrical industry. While their skills tied the central office men to the company, they also won special recognition as the company's most valuable work force. These highly skilled workers began to see the union as an obstacle because the company was willing to give them increased pay and benefits. The union, of course, wanted to maintain a craft rate—a single rate of pay for all journeymen—using the power of the strongest to enforce the rights of the weakest.

Initially, the EEO was a loosely structured group of employees basically serving as a conduit to management. But when IBEW 310 folded and the depression in 1929 began to erode wages and working conditions, craft workers previously hostile to the EEO started to join the organization. The new members included many who hoped to change the company union.[14]

The EEO, on incorporation under the provincial Societies Act in 1930, took on formal structures and gained legal status. It became a forum for discussion of the company's cutbacks during the depression, but remained powerless and a far cry from a union. It lost all of the rights that the craft workers had won in past struggles, including the closed shop, seniority, and collective bargaining rights backed by signed contracts and policed by union members. The EEO saw itself as a "society of employees of the telephone company," incorporated for the purpose of "discussing and co-operating with management" on issues affecting the employees. Its goal was to work towards "smooth working" relations and "fair conditions" for the employees. With no power and no desire to make demands of the company, this new organization could only go cap in hand to the company seeking favors and individual reprieves.[15]

The depression had a contradictory effect on the industry. On the one hand, massive unemployment and economic hard times saw working people cut household expenses to the bone. Thousands cancelled their telephone rentals. For the first time in BC Telephone's history, installers spent more time taking out subscribers' telephones than installing new sets. In the first years of the depression BC Telephone lost 12,000 subscribers. Frank Owston, who worked in the company shops at the time, recalls that telephones were brought back by the truckload. Back at the shops, they were "put to the hammer" to recover the platinum contact points; the rest was scrapped. In July, 1932, the company lost an average of 40 subscribers a day. The company passed the hard times directly on to the employees in the form of layoffs, pay reductions and cutbacks.[16]

Yet at the same time the industry chalked up significant profits and even expanded. In 1930 BC Telephone profits went over the million dollar mark for the first time. The company expanded its national links by joining with other telephone companies across Canada and constructing an all-Canada telephone link—the TransCanada Telephone System (TCTS). Companies in provinces such as British Columbia, which required major construction to complete the national link, benefited greatly from the surplus of skilled labour and low wages. BC Telephone's share of the network meant the construction of 680 miles of lines, which required 35,000 new poles and 1.2 million pounds of copper wire, at a total cost of close to $1.25 million. But the investment was a sound one. Before the operation of TCTS, BC Telephone lost most of the long distance revenues to the United States, through which out-of-province calls were generally routed.[17]

The TCTS construction work was done by heavy gangs. These

were teams of outside workers—"boomer linemen," a few apprentices, groundmen and a driver—specialized in the arduous line construction. The size of the gangs varied, but usually consisted of a dozen workers, depending on the type of line construction. After the new line construction, major repair or rerouting was completed, the gangs were disbanded, with the workers moving on to seek new work. A major company like BC Telephone, though, employed a number of heavy gangs permanently because of constant upgrading and replacement of thousands of miles of outside plant each year.

In the first year of the depression the only workers affected by cutbacks were the temporary employees. This included most workers with less than two years with the company. Late in 1930 BC Telephone started to lay off permanent craft workers, beginning with those with less than five years, then moving to those with less than ten years. The company explained that layoffs were necessary because of the economic downturn and that seniority would be a consideration, but not the sole consideration. At first the outside workers in the cities—installers, troubleshooters and line workers—were the hardest hit. But layoffs quickly spread to the central offices and inside workers. Workers and work were shifted

Laying the TransCanada Telephone System cable during the depression.

Vancouver City Archives

from one department to another, allowing the company the pretence of abiding by seniority while actually conducting layoffs at its whim. Branches of the EEO wrote letters to management asking for an explanation, but this was the extent of their opposition.[18]

As a lineman, Art Marsden was told that he could either take an indefinite layoff or move to Vancouver Island and join a heavy gang. As a single worker, he chose to move to the island, where he worked through the depression. But married or older employees could not be as flexible and were reluctant to accept the hard and isolated heavy gang work.

Along with layoffs, the company introduced a series of rollbacks for the remaining work force. In early 1931, hours were cut back for craft employees with less than ten years' seniority, first in Vancouver and then in other centres. The work week was cut, from five and a half days to four days, with a corresponding drop in pay. Some workers were cut back to two days, only to be laid off with the next wave of staff cuts. As conditions worsened, more and more workers saw their hours reduced and their wages shrink. In the spring of 1932 the company proposed a 5 per cent wage reduction, which was accepted by the EEO.[19]

By fall the company decided to reduce costs in the Victoria central office by taking advantage of new automatic equipment. The EEO was informed that the company was creating a new classification which would include routine maintenance of the new equipment. Routiners, as the workers in the new classification were called, were to be paid the same wage as groundmen or labourers—$132 a month, compared to the craft worker's $168. The new classification was designed to undermine the position of the central office craft workers. While the EEO recognized that this tactic could spell the end of the craft, it accepted the new classification, temporarily at least.[20]

The first branch of the EEO was formed in Vancouver, with John Sowerby, a repair shop craft worker who had been with the company since 1906, as president. In 1930, Arthur L. Creech, a central office maintenance worker from Vancouver who had been with the company for fifteen years, travelled to Victoria and Nanaimo to help form EEO branches. The EEO did not have a central body for the first few years, though branches exchanged correspondence and minutes. On March 12, 1932, the first convention of the three branches was held in Victoria. This marked the first organization-wide discussion.

In August, 1930, shortly after the organizing of the Victoria and Nanaimo branches, a memorandum of agreement was signed between the EEO and BC Telephone which acknowledged formal recognition of the organization by the company. At first glance the wage rates of $168 a month for journeymen and $192 for cable splicers appear close to the rates established by the union in the twenties. But this contract included provisions for merit pay, something the company had initiated in 1926 when it broke the union. "Merit" could raise a journeyman's wages as much as $40 over contract rates. The catch was that merit increases were the arbitrary right of management. Two workers with the same experience and skill could receive different rates of pay for the same job, depending on how well they got along with management. This system encouraged workers to compete with each other for the favors of management. In short, it undermined the co-operation and solidarity among workers, and thus their power.

The EEO confined its activities in the first years of the depression to the organizing of social functions and self-help programs. The emphasis was on team sports, dances, smokers and picnics, and the company often funded the organization's various social events; the socials and sports events were usually joint EEO and company efforts. As the depression deepened, a relief committee was formed in each of the EEO branches to provide material help to laid-off co-workers. No other committee of the branches had as rapid a turnover of volunteers as the relief committee, which is some indication of the despair and helplessness felt by workers who tried to help the victims of the depression. Early in 1932, the EEO established a sick benefit fund and a life insurance plan. The organization's approach was a throwback to the previous century, when unions were established as benevolent organizations.

By 1933, there was a long list of recurring problems which the EEO sought to resolve with the company. Heading the list was the 5 per cent wage reduction. In the summer, all three branches requested the company to end the wage cut. In Victoria, dissatisfaction was growing over the routiner's classification. Routiners were reported to have worked full night shifts without the supervision of craft workers. There was also general dissatisfaction over the obvious favoritism shown in awarding merit pay. The organization was not prepared to take a position against merit pay, but it was opposed to the company's complete control over the determination of merit.

After much discussion over pay reduction, the company changed the cut to 2½ per cent. Late in 1934, a new agreement substantially similar to the 1930 agreement was signed. The EEO had unsuccessfully attempted to bring the employees of Canadian (BC) Telephones and Supplies Limited—a subsidiary which repaired and installed central office equipment—into this agree-

Cable splicers used hot wax to seal and dry the connections in a splice.

Vancouver City Archives

ment. Nevertheless the EEO continued to sign up workers at the subsidiary. The organization did succeed in including the workers in the Kootenay, Chilliwack and Mission telephone companies in the agreement. They also eliminated the routiner's classification. The 1934 agreement restored the 1930 wage rates, but a "Supplement Agreement," appended to the main agreement, continued the 2½ per cent reduction. The company guaranteed that, by the first half of 1935, all craft workers would be on a five-day week. While this still was a half-day reduction of the work week, the wage reduction would then be rescinded.[21]

Throughout the thirties the organization was caught in a dilemma partly of its own making. The EEO was unable to hold its own in the early years of the depression. Wage reductions, layoffs and shorter hours were common in many industries. As a company union, the EEO had to rely entirely on the goodwill of the company. The local autonomy of each branch also made it impossible to generate any solidarity among its members, or to present the company with a single, majority view. Member branches, and later even individual workers, cut special deals with the company which aided the individuals in the short run, but

A central office worker testing and repairing a switchboard.

Vancouver City Archives

resulted in an overall decline in working conditions. EEO members were willing to allow discrimination and violation of the collective agreement as long as it did not directly affect their immediate conditions and wages. The company took advantage of these divisions by introducing changes in any branch willing to accept them. It would then push to make the new conditions universal. Eventually the EEO saw closer collaboration between the branches as vital, as was some form of cross-branch executive which would allow the organization to speak with a single voice.[22]

In order to achieve some degree of consolidation, the 1933 convention created an organization-wide executive which would meet throughout the year and represent members in all meetings with the company. Reflecting the EEO's company union orientation, its executive was called the "directorate." The directorate comprised delegates elected from each of the three branches, with Nelson admitted as a branch for the Kootenays in 1936. The general secretary also was a member of the directorate. This officer—chief spokesperson of the organization—was nominated by the directorate and ratified by a vote at general meetings. The general secretary led delegations to the company and was the chief administrative officer of the union.[23]

R. H. (Reg) Milner, a Vancouver craft worker and former president of the influential Vancouver branch, became the first general secretary. Milner was a model of the new telephone worker, typical of many of the leaders of the EEO. He had started with the company in 1913 as a messenger, showing such speed in this job that he won the lasting nickname of "Cyclone." He served in the armed forces during World War I, but returned to the company in 1919 to work as an apprentice switchboard worker. After 1926, Milner worked as a central office man at the major Vancouver exchange on Seymour street. He even married a telephone operator from the Fairmount exchange. Like many who worked for BC Telephone at the time, Milner identified entirely with the company. His work, his social life and his friends were all somehow related to BC Telephone. He viewed problems with the company as the fault of individuals in management. While the interests of the workers and those of the company could at different times appear to conflict, for the most part the interests of the workers were identical to those of management. It could truly be said that the telephone company was his life.[24]

The new structure centralized complaints, but did not substantially change the character of the organization. For the next five years the directorate met with fierce company opposition to its attempts to enforce the agreement. Worse still, it had to contend with special deals cooked up by different branches and individuals. Their weak bargaining position was soon to change.

By the beginning of 1937, the economy had picked up and, for the first time since March, 1931, all craft workers were working a full 44-hour week at full wages. A number of workers were recalled from layoffs. The EEO began publishing a quarterly magazine, the *Telephone Worker of BC*. The organization even began to see itself as part of the labour movement, sending correspondence to the two labour centrals in the province—the Trades and Labour Congress and the All-Canadian Congress of Labour—seeking information on affiliation.[25]

These changes reflected changes that were taking place throughout Canada. Workers were starting to fight back against government and employers' indifference to the depression. Two years earlier, the unemployed who were shut away in the interior in work camps had gone on strike and marched into Vancouver. From there they took their campaign to Ottawa; they organized the On-To-Ottawa Trek, which was brutally broken up by federal police in Regina.

The EEO did not remain aloof from the new spirit among

A lineman.

working people. The EEO supported the west coast longshoremen in an unsuccessful strike/lockout solidarity action in 1935. A speaker from the embattled union addressed the EEO branch meeting in Vancouver, and the EEO contributed to the strike fund. When the woodworkers started an organizing campaign, EEO branches gave donations and invited woodworkers to send speakers to EEO branch meetings. The members were once again becoming aware of the value of a union and were starting to support and identify with other trade unionists.

But the summer of 1937 brought a return to economic slump. General unemployment increased, and in February, 1938, the company gave notice of layoffs in heavy construction. In the first years of the depression some workers had avoided layoffs by travelling to the interior to construct the TransCanada line. They stayed on to help maintain it during winter blowdowns. But with the new slump in 1937, and not much chance of a blowdown in Vancouver, there was no escape from layoffs. The company proposed that men in the repair shop accept a 40-hour week in place of layoffs. The EEO, however, opted to maintain the 44-hour week, as experience had taught workers that the 40-hour week was no guarantee against further layoffs.[26]

The operators survived the depression without layoffs. The few reductions that occurred came about through attrition. In the years preceding the depression the company had estimated that "the equivalent of the entire operating staff changed every four years." But with the hard times there were few resignations and the operators' school closed down. It was not until 1935 that there had been enough resignations to necessitate the reopening of the school. That year there were 1,400 applicants for 122 operator jobs.[27]

The depression ended with the massive mobilization for World War II. While initially there was little change in technology or working conditions, the war eventually was to destroy the complacency of the EEO. The military was anxious to recruit telephone craft workers for the signal corps and even met with the directorate to see if the organization could encourage enlistment. Civil defence groups—air raid patrols—sprung up among employees of BC Telephone, and the company worked closely with the military to provide telephone communications along the west coast for the military.[28]

The Japanese entry into the war changed dramatically the character of the war on the west coast. There was a massive demand for telephone service for coast defence, and the telephone system was integrated into the war effort. Early in 1942, BC Telephone began one of its largest construction projects in co-operation with the federal government. A network of land lines and radio telephone circuits was constructed along the rugged mainland and Vancouver Island coasts. The land lines alone required 17,000 miles of telephone wire. A new extension of BC Telephone's building on Seymour Street served as an emergency headquarters for the Pacific Command, which was located at the Jericho Beach air station. The building, which was windowless, had three-and-a-half-feet-thick walls; the roof was five feet thick and made of reinforcing concrete and steel to protect this vital communications centre against enemy attack.[29]

The productive energies of the nation were harnessed to satisfy the war demands. The federal government passed the National Resources Mobilization Act, which initiated registration of manpower and limited military training for workers. The company, at the EEO's request, gave half pay to those employees who missed work in order to participate in military training. Subscribers were told not to use telephones during blackouts and air raid tests so as to free lines for the military and emergency services. As the company explained, "This is total war, and the telephone is now reserved for doers and fighters, not mere talkers. And let us all bear in mind that the telephone girls are fighters, too—behind the lines!"[30]

In the interior, vigilant police, military, and even local citizens suspicious of anyone travelling with wire cutters or other telephone "tapping equipment," harassed linemen. The Nelson branch of the EEO finally asked the directorate to request that the company issue special identification cards for outside plant employees. In Vancouver the men complained of "fingerprinting of telephone employees when entering military areas and the inconvenience and subjection to 3rd degree methods, whilst in the performance of duties." The war also brought shortages, rationing and another round of conscription. The company designated many of its craft employees as "vital to the war effort" and ineligible for the draft. The EEO wanted to have a say in who received this designation, but the company refused. Though designated as

Equipment was in short supply during the war.

Telephone Talk

necessary war workers, craft workers had difficulty in getting tools, gas and tires—all rationed items. Through the EEO, telephone workers finally established priority in their purchase of these materials.[31]

The greatly increased demands on the telephone network combined with wartime shortages to create a boom for the telephone company. The operating load of the telephone plant was filled to capacity in the coastal region, but there was always more demand for new military and government lines. Subscribers who had lost telephones in the depression sought to re-establish service now that they were again employed. The company explained the shortage in one advertisement: "Even if we had enough telephones (which we haven't), and enough wires (which we haven't) we still wouldn't have enough central office equipment to give service to everyone."[32]

The wartime demand for labour and the re-emergence of the labour movement led to significant changes in industrial relations. The federal government froze wages by establishing "fair wage" rates, at levels which prevailed between 1926 and 1929. Wages could increase slightly through legislated cost-of-living bonuses. Profits of course were not touched.

The most important change for the telephone workers came with the introduction of two major pieces of labour legislation. In 1943, at the request of the labour movement, the provincial Industrial Conciliation and Arbitration Act (ICA) was amended. These amendments provided for compulsory union recognition and discouraged company unionism. A year later, in February, 1944, the federal government passed an order in council which established the right of employees to form and join unions, prohibited unfair labour practices, set up machinery for defining bargaining units and providing certification, required compulsory collective bargaining and conciliation, and established the right to strike once a collective agreement had expired.[33]

The EEO leaders thought that the amended ICA Act would not apply to the EEO because it was a craft organization and the legislation was aimed at stopping company unions. The EEO general secretary, Reg Milner, reported to a Vancouver meeting "that organized labour [through the new legislation] had put a stop to company unions, but we as a craft organization were not affected." The company, however, initiated a new "arm's length" relationship with the EEO. The legislation stipulated that companies could no longer provide financial support to unions. The EEO was notified that the company would have to cancel its free telephone service, along with numerous other gifts that the company had made to the organization.[34]

The legislated separation between the company and its employees' organization had a positive effect on the EEO. The change spurred the organization to become a union in fact and in name. The directorate wrote to the federal government asking for information on how to become registered as a "trade union." But more impressive than the transformation of the EEO were the changes taking place in the traffic department.[35]

The legislation opened the door to organization among the operators. By the spring of 1943, they were organizing their own union. Within months the operators had 950 members signed up out of a possible 1,100 in Vancouver. Victoria reported that 97 per cent of the operators in their territory were organized into the Telephone Operators' Organization of British Columbia.

The operators' organization had a similar structure to the EEO, though the operators termed their leadership body an executive council. Jeanne Cameron, a Vancouver operator, became its general secretary. Jean Davidson was elected chair of the council. She was replaced a year later by Francis Shaw. The EEO supported the new organization and decided to postpone its own change of name and status until the operators' organization was formed so that both organizations could evolve together. The two sectors of the industry were again working together.[36]

The war had done much to renew the prestige of the operators. Operators were portrayed on billboards and in advertisements as important and conscientious war workers. "No one behind the lines is doing a more useful and efficient task than the telephone operator at her station and switchboard," claimed one such tribute. The war had opened a number of relatively well-paying jobs to women—jobs which had never been available before and which paid more than telephone operating. The wartime expansion in the industry created an unprecedented demand for operators. The increased demand and the fact that many other jobs were available created by the mid-war years a shortage of operators. This placed operators in a position to once again make gains in working conditions and security.[37]

Through the summer of 1943 the operators' general secretary, Jeanne Cameron, assisted by her craft counterpart, Milner, met with the company and negotiated a working agreement for the operators. The agreement included five different classifications of offices. Only Vancouver, New Westminster and Victoria were classified as "A" offices, where operators received the highest rates. Operators took five years of incremental increases to work up to full pay, which was determined by the office classification. Top wages in a "C3" office after five years' service was $3.00 a day; in an "A" office it was $3.80. Supervisors were included in the agreement

During and right after World War II BC Telephone experienced a shortage of operators.

Telephone Talk

and received a top daily wage of $4.30 after two years.[38]

The leadership and many of the most active members of the operators' union tended to be women who had no other financial support. Their tenacious attitude toward their jobs was underscored by the fact that the operators included a unique discriminatory clause in their contract which gave single women a better chance at higher-paid supervisory positions. The clause read as follows:

> Married women not on the permanent staff shall not hold a supervisory position, if they have outside support, and there are single girls to fill the position.[39]

A major problem with the operators' agreement was the wide variance in types of employment. Aside from the five different classes of offices, within each office there were permanent employees, part-time employees, temporary permanent employees and temporary employees. There were night shifts, day shifts, broken shifts, holiday shifts, weekend shifts and relief shifts. The variety of shifts allowed the company maximum efficiency and minimum cost in covering the boards during peak traffic periods. The scheduling was an organizer's nightmare. Rarely did workers get to work more than a few shifts together, and small groups of workers arrived and left work every few hours. The complex shift scheduling made policing the working agreement and collective action difficult. It was a system characterized by favoritism, inconsistencies and special deals.

In the spring of 1944, the operators' council and the craft worker's directorate met to create a new organization, the Federation of Telephone Workers of British Columbia (FTW). Its purpose was "To unite and bring all telephone workers of British Columbia into relations of mutual assistance and co-operation." At its head was the executive council—a year later the name was changed to the board of directors—comprising the general secretaries and two councillors elected from each division. The president of the union, who was to chair council meetings and interpret the constitution, was elected by the council. There were no full-time officers, though officers were reimbursed for time loss and expenses. The two divisions—plant and traffic—each had its own governing council responsible to the branches (later locals), which were led by elected executives. The clerical workers were not yet organized, but the FTW had already linked with activists among these workers. The FTW was optimistic that the writing was on the wall and the clerical workers would soon be part of the federation.[40]

The first attempt to organize the clerical workers had been initiated by the operators. Traffic's attempt to include some clerical workers in their organization was rejected by the company on the grounds that clerical workers were outside of traffic jurisdiction. The fact that the FTW would accept the company's determination of appropriate bargaining unit indicates that the organization had not broken with company unionism. Accepting the company's divisional structure, the FTW decided to add a commercial division. The small clerical work force was organized into the commercial division in 1945. Its general secretary was Robert Smith; Grace Mitchell was elected president. They immediately changed the name of their section of the FTW to the clerical division. By February, 1945, membership of the federation stood at 2,056, with 1,305 members in the traffic division, 575 in the plant division, and 176 in the clerical division. There were still many workers to be organized at BC Telephone. The traffic division organized the company cafeteria workers in Vancouver, and the plant division created two general branches—one in Vancouver and one on Vancouver Island—to include non-craft workers such as building maintenance workers, and janitors previously excluded from the EEO.[41]

While the union appeared to be a single union with three divisions, in practice the emphasis was put on autonomy of the divisions—plant, traffic and clerical. The organization was indeed a federation, of three very different groups of workers. The 1944 constitution explicitly stated that the divisions would "remain forever autonomous and be free from interference in the conduct of their internal business." This unique structure of the FTW, with its three autonomous divisions, led both detractors and supporters to nickname it the "three-headed monster."

Chapter 5 *A National Union*

The wave of organizing which started during World War II and led to the creation of the Federation of Telephone Workers continued into the post-war period. Even the Electric Employees Organization had seen the need for a national organization of telephone workers, but it had set aside this task in favor of organizing BC Telephone employees. With the FTW now organized, one of the first items on the agenda was to initiate a meeting of all telephone worker organizations across the country in order to create a Canadian federation. The telephone companies had for years partcipated in a variety of associations which met regularly and shared information. Telephone workers on the other hand had only informal, if any, lines of communication. At bargaining time organizations would sometimes request copies of each others' contracts. Occasionally an organization would request information about operating procedures or company practices in other provinces. A cross-country federation of telephone workers would open formal channels of collaboration. They could meet together and work out strategies for bargaining, aid each other in organizing, pool resources for research and support each other through solidarity actions, donations or loans.

The FTW modelled their concept of a Canadian federation of telephone workers after the United States telephone workers' organizations which had amalgamated during the thirties and forties. In the United States Bell telephone workers had been organized into company unions in the twenties. These organizations were part of management's efforts to block legitimate unionism. But in 1935 President Roosevelt introduced, as part of his New Deal, the National Labor Relations Act. Familiarly known as the Wagner Act, this new legislation defined unfair labour practices, set up the machinery through which unions could gain recognition, and established a National Labor Relations Board to enforce the act. The act initially had little impact because many companies, including Bell, were convinced that the legislation would be overturned by the judiciary. But in 1937 the United States supreme court ruled that the Wagner Act was constitutional, forcing companies to withdraw their support from company unions. With the aid of the "hands off" provisions of the Wagner Act, company unions at various Bell subsidiaries in the United States achieved wonders. They transformed themselves from company unions to legitimate labour organizations. They established a loose confederation of autonomous unions—the National Federation of Telephone Workers. But after an unsuccessful strike in 1947 the body became more centralized. It reorganized on an industry-wide basis by combining the three separate unions which had existed among plant, traffic and clerical workers into a new single union—the Communications Workers of America (CWA). Two years later the CWA voted to affiliate with the Congress of Industrial Organizations (CIO), the industrial labour central.[1]

Throughout 1945 and 1946 the FTW contacted organizations of telephone workers throughout Canada to set up a meeting to discuss a Canadian federation. Organized labour in Canada had grown by leaps and bounds during World War II. With the lifting of many wartime restraints, workers felt ready to draw all telephone workers together in a common organization. There were two large union centrals in Canada, the craft Trades and Labour Congress, where the IBEW held the jurisdiction for telephone workers, and the Canadian Congress of Labour (CCL). The CCL—comprising Canadian locals of the CIO industrial unions and a number of national unions—had a few telephone locals but had not yet given any union jurisdiction over telephone work.

By 1946 the FTW had a mandate from its membership and sufficient positive response from the other organizations to set a date and begin planning for a conference of telephone workers to be held in Vancouver in the first week of November.

Delegates from the three prairie provinces and Ontario attended the conference. Organizations in eastern Canada unable to attend the conference sent supportive letters and cables. The hosting delegation consisted of Robert Williams, president of the FTW, and the three general secretaries, C. William Mellish from plant, Anne McNaughton from traffic and Eric Mallot from clerical. It was fitting that the FTW was hosting the conference. The union was unaffiliated and would not be viewed as seeking to raid unions already affiliated to either of Canada's two labour centrals. The federation also represented workers in all three divisions of the industry. The FTW had the respect of many of the other telephone worker organizations and was acknowledged as having one of the better contracts in the industry.

The FTW particularly wanted two groups of telephone workers to attend the conference—the operators and plant workers at Okanagan Telephone. OK Telephone had begun as an independent telephone company in 1907, incorporated by an act of the legislature. It came under provincial regulation when the Public Utilities Commission was created. The company quickly expanded from its initial Vernon exchange into the nearby communities of Penticton, Kelowna, Summerland, Peachland, Salmon Arm and Revelstoke by buying up a number of smaller local companies. By 1934 OK Telephone had interconnected with BC Telephone, and through this link was able to provide its customers with long distance service to most points in Canada and the United States.

The plant craft workers were the first to organize at OK Telephone, forming the Okanagan Telephone Company Plant Employees' Association in 1938. This was a company union similar to the EEO at BC Telephone. By 1946 the Plant Employees' Association had a membership of 38 members, the entire permanent craft work force. The plant workers suffered from isolation. Their limited resources and their widespread membership meant that they could hold only one general meeting a year. A federation held out the hope that pooled resources would make possible newsletters, newspapers and travelling organizers to alleviate their problems.[2]

The operators at OK Telephone organized during World War II. As with BC Telephone, wartime labour legislation aided the operators in their organizing drive. But what clinched the drive were the wartime restraints which froze wages at low levels while prices continued to climb. The operators received a great deal of assistance in organizing from Dan O'Brien, the CCL regional director of organization, who persuaded the operators to join the industrial union central. In February, 1944, the workers formed the Interior BC Telephone Operators' Union and received a charter from the CCL.[3]

Shortly thereafter the operators submitted a brief to the British

Columbia Regional War Labour Board outlining their wage demands for parity with BC Telephone operators. The request was refused. Instead of parity, the Regional Board awarded a 10 per cent increase, which was lower than an earlier offer made by the company. The operators were furious. They discussed strike action although a strike in wartime was illegal. O'Brien was able to head off strike action by convincing the operators to appeal the decision to the National War Labour Board. He promised that CCL representatives in Ottawa would appear for the union at the appeal. Unfortunately, because of a mix-up and late notice, no CCL representative attended the hearing, and the appeal was lost. The failure to follow through on the operators' case undermined support for the CCL. The union remained in the CCL, but members felt that the labour centre had let them down. However, they recognized the need to link up with a larger group which could assist them in negotiations and organizing. When the FTW proposed a Canadian federation of telephone workers, the OK Telephone operators responded enthusiastically.[4]

The Okanagan operators' delegates to the conference were Evelyne Tebo and Nellie Forbes, the president and secretary treasurer, both founding members of the union. The OK Telephone Plant Employees' Association also sent its president, A. B.

When the end of the war created more demand for telephones than BC Tel could handle, the company bought this billboard to urge patience.

Telephone Talk

Edwards, to attend the conference and report on any proposed amalgamation.

J. H. Bowcott attended the conference as representative of the Bell Plant Employees Association. With 4,900 out of the 5,000 Bell plant workers in Ontario as members, the Bell association represented the largest single block of telephone workers in Canada. The association was in many ways the most company-dominated of the "independent" unions. The company paid the association's full-time officers. General meetings were avoided; instead the association polled members through referendums. In spite of its company orientation, FTW conference organizers welcomed representatives from the association. Workers at Bell would necessarily play a key role in any federation due to their numbers and because of the leading role that Bell played as Canada's largest telephone company in setting the industry standards.[5]

The telephone craft workers in Manitoba were also members of a company union, the Independent Brotherhood of Telephone Workers. Their secretary, Henry Peters, attended the Vancouver conference. The Independent Brotherhood had been established in January, 1920. Sympathetic strikes were banned by the Brotherhood's constitution; a strike vote of 90 per cent was required for other forms of strike action. Like the FTW, the Independent Brotherhood was changing, spurred on by wartime labour militancy. They too were corresponding with other labour organizations and investigating affiliation with the TLC or CCL.[6]

With the exception of representatives from British Columbia, the largest delegation to the conference was the four delegates from Saskatchewan. Their union—the United Telephone Workers of Canada (CCL)—had organized telephone workers at Saskatchewan Government Telephones on an industrial basis. The union had evolved similarly to the FTW and other telephone worker associations in Canada. But in 1944, Saskatchewan elected a Co-operative Commonwealth Federation (CCF) government. The government immediately enacted the Trade Union Act, which provided for recognition of unions and gave government employees the right to bargain collectively. The telephone workers were among the first to seek recognition and bargaining rights, thereby transforming their association into a union. They also linked up with one of the labour centrals, selecting affiliation to the CCL because it allowed the telephone workers to remain as a single industrial unit. The Trades and Labour Congress required separate locals for plant, traffic and clerical workers.[7]

The last group at the conference was the International Brotherhood of Electrical Workers. Fred Keeley, the IBEW international representative for western Canada, attended the conference at the request of the Calgary craft workers at Alberta Government Telephone. The IBEW held jurisdiction within the TLC for telephone work and hoped to persuade the telephone workers to affiliate directly to the TLC or alternatively to include craft and traffic workers in separate locals of the IBEW. The IBEW, as a craft organization, was not interested in retaining the clerical workers. As part of the courtship of the telephone workers, TLC president Percy Bengough sent fraternal greetings and good wishes to the delegates attending the conference. He assured them that moves to affiliate would be welcomed by the congress. Keeley made it clear from the beginning that he was not empowered to make any final decision but that he would be reporting back to his members and the IBEW on the various proposals which came out of the conference.[8]

It was a diverse group which met at the Vancouver conference, although all delegates were officers of telephone worker organizations. The FTW hoped to forge an amalgamation, leaving the new Canadian federation to decide whether to remain independent or to affiliate to one of the labour centrals. The conference quickly focussed on the issue of amalgamation versus affiliation. Representatives of the two affiliated organizations, the IBEW (TLC) and the Saskatchewan telephone workers (CCL) viewed amalgamation by itself as a step backwards into company unionism. They shared the hopes of the FTW, but their vision was of a single union of telephone workers, with "union" the operative term. An independent federation would be nothing more than one large company union.

Early in the conference Cy Palmer, president of the Saskatchewan telephone workers, criticized the FTW for not "putting to its membership" the issue of affiliation to the CCL. Keeley from the IBEW followed Palmer, noting the FTW's refusal to take action on a TLC affiliation proposal. Anne McNaughton, traffic general secretary, defended the FTW's position by explaining that the decision not to vote on affiliation to any labour central was based on their belief that such a vote would split the FTW membership less than a year after BC Telephone workers had finally achieved unity. Mellish, the plant general secretary, summarized the FTW's position: "While we are anxious to get an eventual affiliation with an established trade union movement, we felt that previous to that it was necessary to get all the telephone workers together, then affiliate as a body."[9]

Delegates from the Okanagan, Manitoba and Ontario—all independent unions—supported the FTW in its stand for amalgamation first. The Okanagan operators were already affiliated to the CCL, but they recognized that their small size and the absence

Laying cable under Victoria. The post-war boom saw unprecedented growth in the industry. The man in the hole is Bert Johns, later general secretary of the plant division.

Telephone Talk

of links to other telephone workers limited their effectiveness. Belonging to a large labour central was not enough, they wanted to be part of a broad organization of telephone workers. Few Manitoba telephone workers were organized and those that were organized were still divided by type of work and geography, so amalgamation was seen as an important step in bringing Manitoba telephone workers together. The Ontario delegate feared that affiliation would be unacceptable to his membership. The Bell association was not interested in affiliating to a labour central, possibly because this would spell the end of their company-dominated regime.

To everyone's surprise a compromise was hammered out. It satisfied the desire of the independents for amalgamation, while it placated the CCL and TLC. Amalgamation would go ahead with the issue of affiliation high on the agenda of the new federation—the Canadian Federation of Telephone Workers, A draft constitution was drawn up for referral to the members of the organizations attending the conference as well as any other organization of telephone workers. A subsequent conference was set for Kenora, Ontario, a year later. The eastern location was chosen in order to include delegates from the Maritimes.[10]

The proposed constitution included a declaration of four principles, the first of which read, "The first objective principle of the Canadian Federation of Telephone Workers shall be affiliation with the trade union movement as an initial step." The other principles were autonomy of member unions as long as autonomy did not conflict with the federation's principles, a commitment to standardized agreements on working conditions and pay, and the promotion of "good Labour-Management relations."[11]

The proposed constitution was a remarkable document which, like most compromises, offered something for everyone. For Bowcott and others still heavily influenced by the company there was the position on good labour-management relations. For both the TLC and the CCL affiliation was made an "objective principle" and slated as "an initial step." For telephone workers in general there was the prospect of standardized pay and conditions. The FTW felt that the proposed constitution laid the basis for a cross-country organization of telephone workers, a step which would reap great benefits.

In the months following the conference the FTW called mass membership meetings around the province. The membership received reports on the conference and voted on the proposed constitution. By the end of February, 1947, locals in all three divisions had endorsed the proposed federation and given the leadership a mandate to continue working toward amalgamation.

But by the late spring it became clear that the FTW was alone. The other unions took no concrete steps towards amalgamation. The Ontario organization made no arrangements for the forthcoming conference and did not even report back to its membership on the proposed amalgamation. The IBEW in Alberta too did nothing, failing even to respond to FTW correspondence. From Saskatchewan the United Telephone Workers periodically inquired to see if steps had been taken to implement what they referred to as the federation's first principle—affiliation. Clearly they had not moved from their position of affiliation first.

The FTW had no alternative but to cancel the conference. In its place they proposed a conference to be held in Manitoba in the fall. After several months it became apparent that only British Columbia and Manitoba delegates would attend, so it too was cancelled. The United Telephone Workers then proposed a conference to discuss affiliation to the CCL, but the FTW rejected this suggestion as it still regarded amalgamation as the preferable route. The unity which had been achieved briefly at the 1946 conference had given rise to false hopes. The divisions that separated Canadian labour were too deep for telephone workers to ignore. There could be no amalgamation without affiliation; and affiliation would mean choosing between the TLC and the CCL—as neither the craft TLC nor the industrial CCL would permit member unions to be affiliated with the other congress. Either way, some telephone workers would be excluded.

The Okanagan workers were still anxious to join with the FTW, and Manitoba had received the go-ahead to proceed wth amalgamation. But both the British Columbia and Manitoba workers recognized that the huge distance separating the two organizations would be a major drawback. After the failure to achieve a Canadian federation, the FTW again held mass meetings around the province to free the leadership from its commitment to seek national amalgamation. The FTW decided to consolidate the provincial telephone workers into a single federation. The traffic and plant divisions of the FTW approached operators and plant workers from the Okanagan and asked them to join the federation. The OK Telephone operators returned their charter to the CCL and in the fall of 1948 became Traffic Local 16 of the FTW. The plant workers joined a year later, becoming Plant Local 6.[12]

In the summer of 1948 the Communications Workers of America had suggested that the FTW affiliate. The CWA—formally the National Federation of Telephone Workers—had just voted to join the CIO. This meant that they were given the jurisdiction for telephone work from the CIO and its sister organization in Canada, the CCL. The main attraction to the FTW leadership was that the CWA was a large industrial union of telephone workers. The following year the FTW board of directors met with representatives of the CWA to discuss the CWA's constitution and what help the American union was willing to give to the FTW in its efforts to create a national federation. Though the FTW leaders knew that affiliation to the CWA would automatically exclude telephone workers affiliated to the rival craft congress—the telephone workers in the IBEW—they decided to endorse affiliation to the CWA, authorizing a fall referendum on the issue.

The September issue of the *Federation News and Views*—the quarterly publication of the FTW—was devoted entirely to promoting affiliation to the CWA. Under the general slogan "In unity there is strength," the bulletin explained what the CWA and CIO were, why the board supported affiliation, and ran a series of questions and answers on affiliation. The plea ended with the slogans, "Let's be constructive, not destructive" and "Let's build a telephone union for all telephone workers."[13]

Opposition was not long in coming. The Victoria plant local condemned the federation leadership and published three bulletins against affiliation. The local pointed out that the CWA was not an international union (though it would become one if the FTW joined). How then would affiliation to an American union bring about further unity for Canadian telephone workers? The local also argued that it was highly unlikely that the CWA would improve the FTW's bargaining strength as the federation had a better contract than the CWA locals. But the main focus of the bulletins was a denunciation of the leadership for using the *Federation News and Views* and the funds of the organization to campaign on behalf of a "foreign union."[14]

Extreme in their polemic, the bulletins nevertheless fitted within the bounds of healthy opposition. Less legitimate though was the intervention by the company. Management personnel lobbied FTW officers in opposition to affiliation. In July management met with union officers to express their "disappointment" that the union's board of directors' had endorsed affiliation to the CWA/CIO. Management expressed the view that affiliation would lead to a deterioration of labour relations at BC Telephone. They raised the possibility that the CWA might revoke the FTW charter if members later voted to disaffiliate. Finally management warned against possible "involvement of the Federation in a sympathetic strike."[15]

The BC Telephone vice-president and general manager, C. B. Diplock, wrote a sharply-worded letter to the federation secretary treasurer, Don Weaver, stating that while he did not wish to interfere in the union's internal affairs, the company felt that it should draw the union's attention to "certain vital information." Diplock maintained that affiliation would be a "repudiation by the federation of the excellent employee-management relations which have prevailed for the past quarter of a century." Further, it would open the federation to "political domination from without," giving the CWA power to "freeze [federation] funds," "revoke the federation's charter," intervene in bargaining with the company, and possibly "prescribe additional assessments." The company condemned the "attitude of militancy" which accompanied the pro-affiliation arguments and suggested that such militancy was "foreign" to management-union relations at BC Telephone. Finally Diplock noted that the CIO "has a history in which work cessations are relatively frequent." The company made sure that its views against affiliation were known to the FTW membership; copies of the letter were circulated to supervisors who were advised to make the company's position known to the work force.[16]

Throughout November meetings were held around the province, with locals voting on a referendum worded, "Are you in favour of joining the CWA/CIO?" The ballot was a disappointment to the board of directors. The only division to favor affiliation was traffic. Only 856 operators out of a possible 1,700 voted, with 506 for and 350 against affailiation. The plant division, with 900 members, led the campaign against affiliation, with a vote of 344 for versus 480 against. The tiny clerical division also went against affiliation, voting 105 for versus 153 against. The total federation vote was 955 for versus 983 against affiliating. The movement to join the CWA was defeated by a combination of nationalist concerns of the membership and company interference. The ease with which the company participated in this internal union debate, and its success, indicated that the federation was still heavily under the influence of the company. The operators—the largest group in support of affiliation—never contemplated affiliating their division alone, preferring instead to reopen the issue a few years later.[17]

Because of the close vote, the issue was far from resolved. The IBEW approached the federation offering affiliation to the TLC through the IBEW. The federation leadership invited both the IBEW and the CWA to speak to members at meetings throughout the province. This second time around, with both the IBEW and CWA sending representatives to argue the case in favor of affiliation, and much discussion in the months leading up to the vote, led to increased support for affiliation. The membership voted on the question, "Do you wish to join an international union?" and gave a majority vote in favor—1,453 for versus 1,030 against—but the FTW constitution required a two-thirds majority. Once again affiliation was defeated.[18]

One more vote was held, over the opposition of the operators who asked that the issue be dropped. In this final ballot in 1953 the motion was defeated with an abysmal turnout of less than 500 members—the affiliation issue had died, for the time being at least. The issue was closed in January, 1954, when the board adopted a recommendation from both the traffic and plant divisions to drop the matter.

The 1946 conference and the subsequent attempts at amalgamation/affiliation marked a watershed for the FTW. The union had evolved from a company union of craft workers to an industrial union seeking links with the Canadian and United States labour movement. Having failed to create a national federation or link up with an international union, the FTW returned to its internal affairs and attempted to stabilize its position as the largest independent union in the province and at the same time cope with significant changes in the industry. In spite of its failure, the FTW's attempts to form a national federation provided valuable experience. During the discussions on amalgamation and affiliation, FTW officers met with many trade unionists; telephone workers saw different trade unions in action, read their contracts, and became familiar with their operating methods. Contact with other unions taught the FTW what was good about their own organization and pointed out improvements which could be made.

Few unions had a structure similar to that of the FTW with its

Vancouver City Archives

The traffic division in the late 1950s.

three autonomous divisions. Lamenting the structural problems inherent in the FTW's "three-headed monster" became a popular union pastime. But each division jealously guarded its autonomy whenever discussions were initiated on alternative structures. The plant division regarded its members as the only true trade unionists. As late as the fifties the plant division still had members who had worked at BC Telephone when the IBEW had the closed shop agreement. The self-assured, skilled craft workers dominated the plant division and kept local meetings lively. The autonomous structure of the FTW suited the division because they could negotiate their own appendix to the contract and run their own affairs without being dominated by the numerically larger traffic division. Plant felt that divisional autonomy gave it the best of both worlds, a union of all telephone employees in which plant could play dominant part.

For the traffic division, autonomy ensured operators control of their own affairs. While traffic was the largest division, high job turnover, odd shift patterns, and general lack of trade union consciousness made it difficult to marshall the division. Operators feared that an integrated union would mean token representation of operators on an executive dominated by craft workers.

The clerical division—the smallest of the three divisions until the late fifties—also was satisfied with the FTW structure, which guaranteed continued clerical autonomy with loose ties to an industry-wide union.

The most significant aspect of autonomy was its effect on bargaining. The first contracts between BC Telephone and the FTW were negotiated separately by the three divisions. By 1947 there were separate plant, traffic, and clerical contracts, each signed by the general secretary of the particular division. Since 1948 the FTW had negotiated a master agreement which dealt with common conditions; in addition each division negotiated a separate appendix. The master portion of the contract was usually brief, as only a few general clauses applied to all three divisions. The bulk of the contract specifying wages and working conditions lay in the appendices. While separate appendices seemed a logical extension of the union's autonomous structure, they created problems. Wide variations in wages, hours of work, and working conditions existed among the three divisions. Workers in one division were generally ignorant of the problems and concerns of those in the other divisions. This made it difficult for the union to form cross-divisional links and present a common front to the company. Often the company would introduce a seemingly insignificant practice in one division with little opposition. The implications of the practice would only become clear when it was extended to the remaining divisions. Separate negotiations dispersed the union's energies; each division gave different weight to different issues. To appreciate the diversity in priorities, we need only look at some points included in the 1948 agreement and its three appendices.

The 1948 agreement divided clerical workers into ten wage groups. The salaries within a single wage group were graduated into as many as twelve incremental steps; it took more than six years on a job to work up to the top rate. As well, clerical workers in the higher-paid groups, numbers six to ten, had different rates for male and female workers. In the early fifties separate male and female pay rates were eliminated in the clerical division. Unfortunately this overt form of discrimination was replaced by practices which continued the discrimination in a more discreet manner.[19]

Operator wage rates were also heavily graduated; it took an operator six years to gain the full rate. Telephone offices were divided into "A," "B," and "C" categories based on the amount of traffic handled. Lower rates were paid to operators in "B" and "C" offices. Operators worked shifts and received a small shift differential. A shift could be six, seven, or eight hours, depending on the type of office and time of day. Work was scheduled in irregular patterns, for example, ten days on, four days off, followed by six days working with one day off. The traffic division also had a large "temporary" staff which allowed the company to expand and contract its operating work force. The union's acceptance of the classification permitted the company to isolate this particularly vulnerable part of the work force. This enabled the company to institute layoffs while boasting that none of the permanent staff had lost their jobs.[20]

In the 1948 agreement the plant division appendix was divided into four sections: crafts, building services, automotive maintenance and stores. In the largest sub-group, craft, there were four wage groups, each with a number of incremental steps. Group 1 included installers and linemen; group 2, central office men and PBX installers and repairmen; group 3, cable splicers; and group 4, cable splicers' helpers and heavy construction groundmen. The appendix included special wage differentials, board and lodging provisions, and overtime and vacation clauses. The plant appendix also included a clause which forbade contracting out. This clause, part of the plant appendix and not the master agreement, only limited the company's ability to contract out craft work.

In spite of the diversity in the three appendices, there were common issues. Topping the list of priorities at every set of negotiations was union security and the closed shop. The company refused to move beyond simple recognition of the union, and the union did little more than raise the issue each time. The union negotiated a seniority clause which was not effective as it stated only that seniority would be given every consideration. Another perennial issue in negotiations was the reduction of worktime and improved vacation benefits. In 1946 workers in the clerical and plant division won the 40-hour week with no reduction in pay. Two years later this was further reduced by the clerical division to the 37½-hour work week. The union also put the issue of union participation in the pensions and sick benefits plan on the table at each set of negotiations, with little success.[21]

The gradual improvement in wages and inroads in other areas gave rise to the belief within the union that the future would offer incremental improvements. Few members saw anything in the growth of the industry during the late forties and early fifties to threaten their confidence in continued progress. Most believed that the relative security and prosperity of the post-war years would continue. The company encouraged their complacency.

Throughout this period BC Telephone grew faster than ever before. Fuelling this growth was a massive backlog of orders which had accumulated during the war, coupled with an unending flood of new orders as British Columbia became the fastest-growing province in Canada. The telephone was now a household necessity and most people had the money for telephone service, which was one of the few services not to have increased in price since the twenties. By 1948 craft workers at BC Telephone were installing three times as many telephones per month as in the pre-war years, and there was still a long waiting list for telephone service.[22]

Concurrent with this phenomenal growth in service was the company's initiation of a major automation program. The conversion of common battery telephones and manual exchanges to dial telephones and automatic switching equipment marked the beginning of a technological revolution in telecommunications which continues today. While automatic switching technology was patented as early as 1891, it was not until the massive growth in telephone demand in the late forties and fifties that BC Telephone began the extensive conversion to automatic switching.[23]

Automatic switching had a number of advantages over the old manual systems. In order to place a call with the common battery manual system, a subscriber would pick up the receiver. This sent an electrical signal which sounded a buzzer and lit a small indicator light on a switchboard at the central office. The operator, alerted by the buzzer, connected a cord to the subscriber's jack and asked, "What number please?" Then the operator connected the subscriber with the requested number by connecting the other end of the cord to the jack of the requested number. When the call was completed, the light again came on and the operator pulled out the cord, severing the connection. This system required a large operating staff and was relatively slow, as every call had to be manually connected. Manual operating rooms also took up a great deal of space as every telephone line required a jack. A single board

BC Tel's massive post-war growth meant significant growth in clerical staff, which became part of the FTW immediately after World War II.

Telephone Talk

held only a few hundred jacks and an operator could not handle more than three boards at a time.

By contrast, to place a call with an automatic exchange the subscriber merely had to dial the desired number. This sent electrical impulses to the central office, where switches did the selecting and connecting automatically. Thus automatic exchanges eliminated the operator. They increased the speed of connecting calls and reduced the operating staff by more than 80 per cent.[24]

Dial conversion at BC Telephone began with the cut-over of a few small exchanges in the late twenties. Victoria was the only main centre to have converted by the early years of the depression. Wartime demand for telephone service and the saturation of telephone capacity in Vancouver made it necessary to convert the downtown area to dial in the early years of World War II. Further conversions were not possible then because of wartime shortages of equipment. No more conversions took place until the late forties, when BC Telephone began a massive dial conversion campaign which continued into the fifties.

The automatic exchanges had a contradictory effect on telephone workers. It eventually decimated the operating staff and greatly reduced the ranks of Canadian (BC) Telephone and Supplies (CT and S), the company's construction subsidiary. But during the decade of conversions the work force actually grew. This growth further encouraged the union's post-war complacency. At the start of the conversions in 1946, a labour shortage forced the company to make a "radical departure in policy" and hire "married women up to and including age 35" as operators. The company guaranteed these women employment "for at least a year." The work force almost doubled in the decade after 1947, growing from 3,588 employees to 7,130 by 1957. But there were two distinct types of growth in the work force—one of a temporary nature and the other permanent. The increase in operators' ranks was temporary, reflecting the company's need to run manual and automatic systems in parallel while major centres were converting. Similarly CT and S expanded to meet BC Telephone's massive construction program. Its workers converted existing manual systems and constructed the new automated exchanges. At the same time the clerical work force increased, reflecting the growing administrative and record-keeping needs of a company which more than doubled its telephones in service in a decade of rapid automation.[25]

The dial conversion was the largest and most extensive of the changes taking place in telecommunications. But this was merely the leading edge. Before all exchanges had been converted to dial, direct distance dialing was introduced. The new system further cut operating staff by making it possible for subscribers to make long distance connections without going through an operator. New control consoles in long distance traffic centres greatly increased BC Telephone's ability to monitor staff and call levels. The company boasted that the consoles would ensure that traffic supervisors would no longer "have to depend on their fabled X-ray eyes and other superhuman powers to tell them exactly what's going on in their toll rooms."[26]

In the mid-fifties BC Telephone, along with the other telephone companies in the TransCanada Telephone System, began work on a cross-country telephone and television microwave network. By 1958 a system of 155 stations, approximately 30 miles apart, linked major cities across the country. While designed to augment the extensive open wire line system completed during the depression, the microwave network was a qualitative leap ahead. The microwave system could carry much more information, including television and data signals. This opened new markets and allowed telephone carriers to transmit signals faster and improve transmission quality. The system even included sophisticated self-monitoring devices; microwave stations required significantly less intensive repair and maintenance work than comparable open wire links.[27]

There were also major changes in the clerical area. International Business Machine (IBM) data processing equipment was purchased to mechanize record keeping. In the mid-fifties, BC Telephone, by transferring its records for delivery of telephone directories to IBM punched cards, greatly reduced the time required to compile directory delivery records and print delivery labels. The company also began to transfer records of plant construction and maintenance expenses to accounting machines. By the late fifties the company was proceeding with a major conversion program to transfer subscribers' bills, service orders, and the production of shareholders' dividend cheques to machine processing. In addition to introducing new machinery in its own offices, BC Telephone also marketed the "Electronic Secretary," a telephone answering device. The following company assessment of the virtues of the machine pointed in the direction of further automated office equipment:

> The Electronic Secretary retains many of the better features of the private secretary, plus a few improvements. In addition to being compact and attractive, it is efficient, practical and businesslike. Though lacking the charm and vitality of the real thing, the Electronic Secretary makes up for it with unobstrusiveness and reliability. Besides, it can be turned on or off by merely flicking a switch . . . To keep itself at a peak of efficiency, the Electronic

> Secretary needs no coffee breaks or holidays—just the occasional squirt of oil and a steady supply of electricity.[28]

BC Telephone stressed the speed and accuracy of the new technology. In place of the human touch the company emphasized efficiency gains. Describing the cut-over of the large Bayview exchange to dial, the company wrote: "As Bayview died, the cold, impersonal but efficient hum of the dial tone replaced the warm voices of several generations of girls who had staffed Bayview since November, 1911, when the office handled its first calls." In a separate article entitled "automation + telephony = more people," however, the company assured workers that while "some job dislocation" was unavoidable, the effect of automation on employees would be "minimized through careful planning for years ahead and because of the continuing rapid growth of our business." "Automation," the company explained, meant "improvement and expansion of service, and increasing diversification of job skills and widening of job opportunities." The test of the company's word would not take place until the sixties, when the conversion had been completed.[29]

Telephone Talk

Linemen over Victoria during the war.

Chapter 6 *A Time of Change*

The Federation of Telephone Workers changed radically in the sixties. The organization which emerged at the end of World War II had much in common with its company union predecessor. The province-wide boom and the resulting telephone conversion program in the post-war period created thousands of jobs for young workers in the telephone industry. This had enhanced the older workers' sense of security. But good times at BC Telephone ended abruptly in the sixties. Layoffs, changes in working methods introduced by new owners, a developing awareness among workers of the negative effects of automation, and a growing mood of labour militancy across the country all helped to transform the FTW into a fighting union. By 1969 the overwhelming majority of union members were convinced that only strike action could win concessions from an intransigent company.

The layoffs at BC Telephone in the late fifties and early sixties were a major factor in the transformation of the FTW. The completion of most of the dial conversions, together with the winding down of the construction program involving the transfer of lines from aerial wire to underground cable, allowed the company to drastically reduce its work force. At the end of 1959 the membership in the three divisions totalled 5,491. The plant division had 2,573 members, traffic 1,991, and clerical 927. Four years later membership had declined by 1,334 to a total of 4,148, with 2,097 in plant, 946 in traffic, and 1,105 in clerical. Profits in this same period rose substantially. In 1958 the company's profit was $4,173,071. By 1964 when the company finally started to rehire some of the discarded staff, profits were $11,928,655. The three divisions differed in the extent to which they were affected. They also differed in their response to the changes in the industry.[1]

The FTW felt it could do little about the layoffs. The clerical division published articles on the issue of automation in its divisional bulletin, but the division leadership felt that with lack of concern by the membership and low attendance at meetings, little could be done on the issue. This lack of militancy was exacerbated by the fact that, while the membership of traffic and plant declined in this period, the clerical division continued to grow. The plant division had negotiated a layoff clause in its appendix to the 1960 agreement. Clerical included the clause in 1962 and traffic added it in 1964. But the layoff clause in the three appendices failed to protect workers from job loss; the clause merely spelled out the manner and order in which workers were to be laid off. All three layoff clauses began by stating that "Whenever the economic or [work] force conditions are considered by the Company to warrant the laying-off of regular employees, such [work] force adjustment as it may deem necessary shall be effective . . ." With nominal reference to seniority, the company could lay off workers at will.[2]

An article in the *Plant Division Bulletin* argued that there was an alternative to layoffs, namely, improvement of service. The article noted that in the light of "so much visible necessary outside plant re-construction" the real reason for the layoffs was "a tightening of the purse strings rather than a shortage of work." The bulletin argued for corporate responsibility in a recession. "No one denies the money lender a reasonable return on his investment especially during a rising economy, but the ever-increasing rise in profits does not release him from an obligation to the employees, the actual producers of his wealth, and to the economy of the community in which they live and from which he derives his profits." The article noted that the layoffs, with ample work available, were a nail in the coffin of the "paternalistic company which would take care of its employees."[3]

The Victoria Plant Local 2 recommended that a mass grievance on behalf of the laid off employees be filed. They argued that the company had "promised ten years' work" and that the men joining the company had accepted "low wages, with the expectation of rising to the top rate through service." Legal advice on the proposed grievance forced the union to drop the proposed action. The incident dispelled any illusions about the force of moral argument. As the union reported:

> 1. Verbal assurance of 10 years' work—any contract over one year must be in writing. Production of an affidavit has no value only when produced in court and in this case was valueless as the oral contract concerned, if there was one, was for more than one year.
>
> 2. The company's propaganda booklet, received by all employees, was termed by Mr. Johnson (the lawyer) as "puff" and was merely a type of employee encouragement used by most companies.
>
> 3. Moral responsibility—of an employer to an employee—has no legal status whatsoever.
>
> 4. The maintaining of high profit level ahead of maintaining a high employment level also has no legal status and is something all unions are faced with and is material that might be of value in negotiations only.
>
> 5. The only possibility to avert layoffs is through a guaranteed wage or no layoff clause during the life of the contract.[4]

While traffic suffered the largest decline in membership, the cuts were made primarily through attrition. The plant division incurred most of the layoffs. These centred mainly in the construction and CT and S work force. The layoffs shook many of the plant members out of their complacency. Gradual progress had been arrested, and workers again faced possible job loss. Young men attracted by secure employment and willing to accept a low starting rate for the opportunity to get into the craft were shaken with each successive layoff. Don Bremner, an apprentice at the time, recalls seeing co-workers with up to five years' seniority laid off. By 1960 Bremner had been with the company for five years and was well aware that workers with the same seniority were being laid off. Many other workers were demoted to entry level jobs. Larry Armstrong, who worked on a conversion crew, recalls that on completion of the conversion a number of co-workers were given the choice of returning to "the ditch" for a few weeks of construction work or being laid off.[5]

At first the layoffs tended to divide the union into older and younger workers. In the plant division the massive recruitment campaign of the fifties had attracted a new type of telephone worker. According to Bert Johns, the plant general secretary from 1963 to 1970, the workers who joined in the fifties and sixties were "younger and tended to be more militant;" they were not "married to the job." They had come to the company seeking good jobs and expected the union to protect them. Many older workers defended the union, which had helped establish some of the best wages and working conditions in the industry in North America, but the younger workers considered the FTW to be an ineffective company union.[6]

In the early sixties Local 1, the large Vancouver plant local, accounted for close to half of the total membership of the division. It was within this local that militant dissidents began to organize. They argued that the FTW was ineffective and a company union, that its structure led to membership apathy and that Norm Pettigrew, the general secretary and on full-time union staff, was a dictator. Pettigrew, general secretary since 1952, had been re-elected annually by the plant council with little or no opposition. In the fifties, the three divisions placed their general secretaries on full-time staff. With the crisis in the early sixties, the plant division added further full-time union staff—assistant general secretaries. These were essentially business agents who worked with and were assigned work by the general secretary. Union staff in all three divisions were elected by the divisional councils. The dissidents claimed that the plant council, which had one or more councillors from each local, was not representative of the general membership and should not elect the general secretary. They further argued that the union did little in the way of education and proposed that Local 1 should have a newsletter. In the fall of 1962 Local 1 agreed to launch and pay for a newspaper.[7]

The first edition of *Short Circuit* appeared in September, 1962, as the "official monthly publication of Local 1." The paper was actually written and edited by a volunteer staff which did not report directly to the local executive. The first editorial assured readers that the newspaper would "stir things up." The paper did not take long to deliver on this promise. The next seven issues included calls for joining either the IBEW or CWA, attacks on the bargaining team, long treatises on what was wrong with the FTW, and attacks on Pettigrew and his fellow general secretaries from clerical and traffic, Gravel and Lillquist. Each edition became the focus of a major debate within Local 1. Eventually the business of the local was reduced to explosive debates in *Short Circuit*.

Pettigrew was particularly frustrated by the attacks and took the issue to the plant council, which gave instructions to Local 1 to take over control of the newspaper, but to no avail. Pettigrew had been planning to retire as plant general secretary and the opposition in Local 1 might have hurried his departure. In the spring of 1963, at a special meeting of the plant council called to discuss his resignation, the local was ordered to stop printing *Short Circuit*. The council condemned the newspaper for not fulfilling the needs of a union newspaper, but seeking only to "attack and destroy."[8]

Two months later, at the June plant council meeting, Local 1 was placed under trusteeship on the resignation of a majority of the local executive officers after heavy criticism by the dissidents for accepting the council's directive to stop publishing *Short Circuit*. When it was revealed at the council meeting that the dissidents had

Conversion to dial eliminated hundreds of operators' jobs.

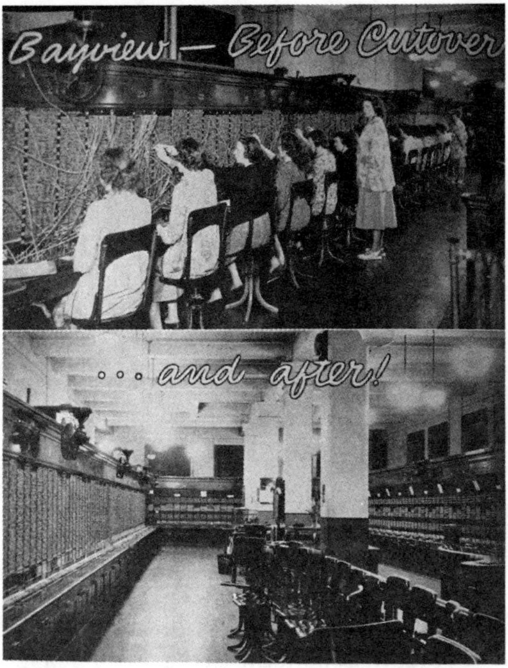

Telephone Talk

held a secret caucus to hatch a plan to force some of the resignations, eight officers and rank and file members of Local 1 were charged as a body with violating their oath of membership—to "further the interests of the Federation at all times."

The members were tried by the board of directors, with newly-elected plant general secretary Bert Johns chairing, but not voting on, the procedings. The eight members were found guilty and given indefinite suspensions. The suspensions were overturned by the board a few months later, after the eight filed a challenge of the procedings in court. The federation's lawyers advised the union that an ambiguity in the FTW constitution made it questionable whether the eight members could be tried as "a body" and that the union should reverse the decision and charge the members as individuals. With *Short Circuit* gone, a new plant general secretary, and Local 1 functioning again, the division decided not to retry the dissidents.[9]

The actions around *Short Circuit* and its supporters disrupted Local 1 and the entire plant division for over a year. A number of plant workers were sympathetic to some of their criticisms, but many felt that the group's constant attack without offering alternatives was destructive. Few defended the group. Robert

Baillie, a telephone worker since 1929, was one of the councillors from Local 1 who resigned in the dispute. While Baillie was critical of Pettigrew's "one-man show" and acknowledged many failings, he felt that the group around *Short Circuit* was "disruptive" and "underhanded." His viewpoint was that they indulged in too many personal attacks and were going about trying to change the union "in just the wrong way."[10]

For Bert Johns, elected plant general secretary at the stormy June council meeting, the trial of the eight was an inauspicious beginning to his term as plant general secretary. Originally from London, England, Johns was very much a "telephone man's telephone man." He had worked as a boomer lineman in Saskatchewan and Alberta on both power and telephone lines, joining BC Telephone on his release from the army in 1944. As an activist in Victoria Local 2, Johns identified with the younger, more militant telephone workers. He criticized Pettigrew for "essentially taking whatever the company offered." But Johns did not hesitate to step into the breach when illness prevented FTW president Hazel Magee from chairing the trial of the eight. For Johns the important issue in the dispute was loyalty to the union. Ironically, while all eyes focussed on the fuss in Local 1, a more serious challenge to the federation, a raid by the Communications Workers of America, was in progress.[11]

Contact between the CWA and FTW had been renewed in the late fifties when the CWA initiated two cross-country conferences of telephone workers. The merger of the TLC and CCL into the Canadian Labour Congress (CLC) in 1956 paved the way for discussions on unity. But long-standing differences over structure and affiliation could not be resolved. The CWA nevertheless maintained contact with the FTW. It paid the expenses for the general secretaries to attend a week-long CWA training conference in Ontario and later footed the bill for two educational seminars for FTW councillors and senior officers in the fall of 1962. Officers of all three divisions were noncommittal about the CWA's courtship, until it became obvious that the traffic division was seriously considering joining the CWA.[12]

In the late fifties and early sixties traffic went through the greatest change of any of the divisions. In a few short years automation decimated the membership of the division. Traffic declined from being the largest division and largest occupation group in the telephone industry to the smallest division and a threatened species in the industry. As with the plant division, the post-war growth had attracted a new, younger work force. Before the general introduction of automated exchanges, manual operators in the various exchanges around the province constituted the majority within the division. There was also a small toll or long distance operating work force. But as the company used automation to close more and more offices, older operators with seniority transferred to the few remaining manual offices, leaving the younger workers concentrated in the expanding areas of toll and directory assistance.[13]

In the late fifties a group of young toll operators in the largest local in the division, Vancouver Local 10, started to agitate for a change. Elizabeth Lillquist was a leader of this group. She was a toll supervisor, a position which was part of the bargaining unit at the time. Lillquist had joined the company in the mid-fifties. Before she started at BC Telephone, she had heard from a number of trade unionists that the FTW was a "company union." In spite of this forewarning, she was still "appalled" to see just how ineffective the organization was. The union had permitted the company to displace hundreds of operator jobs — mainly because the majority of these operators were temporary workers. The traffic division had no educational program, few shop stewards, and rarely processed grievances. In fact it had never taken a grievance to arbitration. According to Lillquist, the Local 10 operators who sided with her felt that the union was run by "people who had worked forever with the company" and spent more time explaining "how you have to understand the company's point of view" than developing the union's position.[14]

In 1958 Local 10 passed a motion to recall Thelma Mathias, general secretary of traffic since 1951. Mathias was well-liked and supported by the operators in locals outside of Vancouver, but Local 10 accounted for close to half the total membership of the division. Mathias resigned and Lillquist, despite opposition by many of the locals outside of Vancouver, was elected by the council to replace Mathias. The removal of Mathias and the election of Lillquist created a schism within traffic between Vancouver and the outlying locals. A number of councillors and locals protested the action of Local 10 and the fact that one local could remove the general secretary.[15]

Lillquist felt that the traffic division needed to align itself with a "larger union" which could give it "more support." She also felt that the "autonomous structure" of the divisions gave traffic division the worst of both worlds. The three divisions lacked the power of a united movement in negotiations, but because the union was certified as a single bargaining unit, the traffic division was legally bound to the FTW. As early as the summer of 1959 Lillquist argued for separate certifications for each of the three divisions. Although she was unable to win the board of directors to this position, Lillquist was nevertheless authorized by the traffic council to investigate other telephone worker organizations, including the Telephone Employees' Association (Bell) and the CWA.[16]

Problems arising out of the 1962 contract negotiations strengthened the resolve of a number of the traffic division leaders to abandon the FTW. In the negotiations the operators had decided to make the shorter work day with no reduction in pay their priority in bargaining. They felt that the seven-hour day would cushion the effects of automation by increasing the number of operating jobs. Neither the plant nor the clerical division saw the demand as a priority. So traffic fought for it on their own, in the negotiations on their appendix and into conciliation. The CWA, meeting in convention in Milwaukee while the issue was still before a conciliation officer, passed a motion offering "full and complete support" to the operators in their fight for the seven-hour day. The operators finally won the seven-hour day and a 1½ per cent increase in pay. Clerical and plant, working a seven-and-a-half and

Changeover required maintaining both old and new systems until new was almost entirely in place, resulting in a significant, but temporary, increase in staff.

Dial and Manual Telephone Systems
NOW WORK TOGETHER IN VANCOUVER

Shelves of switches in the new Marine telephone office. Switches are the devices that handle telephone calls under the dial system.

When a Marine subscriber dials O for operator, he is connected with this switchboard, the **Dial Service "A" Board** in the Seymour office.

A Call Indicator position on the Trinity switchboard. When a manual number is dialed by a Marine subscriber, it appears illuminated on the switchboard in the manual office. The operator completes the connection.

Miss Agnes Galbraith, Trinity operator, dials a Marine number for a Trinity subscriber. Calls to Marine numbers from all manual offices, except Seymour, are handled in this way.

Instead of dialing Marine numbers in the Seymour office, "A" operators pass the numbers to a **Cordless "B" Board** where operators complete the calls by depressing keys.

Another view of the **Cordless "B" Board**. Miss Freda Pearce, instructress, is standing at the left, and the operators are Miss Nina Dempsey, Miss Ethel Gulliford, and Miss Eva Hancock.

Telephone Talk

eight-hour day respectively, received 3 per cent increases. Traffic hailed the seven-hour day as a major breakthrough; they became the first operators in North America to win this shorter work day. But a number of clerical council members advised operators to vote against the shorter day on the grounds that it was not worth the smaller increase and that with shorter hours operators would remain behind the rest of the union in wages.[17]

Shortly after the new agreement was signed, the CWA gave a series of two-day educational courses for FTW leaders which provided the CWA with an excellent forum to launch its raid. Close to 200 leaders from the three divisions attended the courses in Nanaimo and Vancouver. The CWA could assess who would be on their side and who would oppose the raid. William Dunn, CWA District 10 director and organizer of the schools, boasted openly that the "technique of using the standard CWA two-day stewards education school in organizing has proved very successful." The CWA leaders felt that they were not "raiding" another union, but "organizing the BC Telephone workers." In fairness to the CWA, many leaders of the FTW were receptive to being "organized." Norman Gravel, the flamboyant general secretary of the clerical division, wrote in the division's bulletin in the fall of 1962 "that if we, by choice or through circumstance, ever seek the security of membership in a larger organization, the CWA is the logical road to travel."[18]

At the January, 1963, traffic council meeting the operators took the initiative. The council authorized the general secretary to investigate joining the CWA and to arrange for Dunn to address a series of informational talks on the CWA at traffic local meetings around the province. Dunn sought advice from a CWA lawyer and the traffic division lawyer, Tom Berger, on whether it would be advisable for the division to disaffiliate immediately from the federation or seek certification first. The lawyers advised him to seek certification first.

By spring the CWA had opened an office two blocks from the FTW office on Granville Street. They started to publish a bimonthly newsletter — the *BC Telephone Organizing Committee Newsletter.* Throughout the spring and summer, Dunn, along with John J. Moran, a retired CWA vice-president, and Bert Renaud, an international representative, travelled the province arguing the case for the CWA. As late as July Norm Gravel and some clerical councillors sat in on the CWA "educational sessions" as the CWA had not yet brought out its membership cards nor had it officially started the raid.[19]

The raid was formally launched on September 13, when a special meeting of the traffic council was called to explain the recent reversal of the trial of the Local 1 dissidents. After a brief summary of events around the *Short Circuit* incident, the meeting was devoted to the CWA. Berger, as the traffic division legal adviser, was invited to address the council on the problems of affiliating to the CWA. A motion was passed in favor of joining the CWA, followed by a motion authorizing the general secretary to seek a separate certification. When the council meeting adjourned, all councillors but one, FTW president Hazel Magee, signed up with the CWA. A CWA charter had already been assigned to Local C-14. A number of the officers of the newly-created New Westminster FTW traffic Local 14 became the founding officers of the CWA local.[20]

For three months CWA organizers, aided by traffic general secretary Lillquist, travelled to outlying traffic locals and signed operators into the CWA. But the FTW soon mobilized a counter-campaign. Phyliss Bradshaw, an operator and former FTW president, covered the province with Hazel Magee to defend the FTW. Meanwhile clerical general secretary Norm Gravel started another "raid" of sorts on behalf of the clerical division. In an effort known as the "pink card campaign," traffic division members were encouraged to sign "clerical division show of interest" cards, indicating interest in a merger between the traffic and clerical divisions of the FTW. As most senior officers of the traffic division had joined the CWA, Gravel argued that the "pink card campaign" allowed operators the option of staying with the FTW. This campaign gave Gravel and clerical workers the opportunity to address operators on the benefits of membership in the FTW. The FTW leaders realized that though the CWA campaign was aimed at traffic, if it succeeded the clerical and plant divisions would be next in line. At the same time it was clear that the removal of the traffic division from the federation would spell the eventual demise of the FTW.[21]

For four months charges and countercharges flew back and forth between supporters of the FTW and the CWA. The CWA charged that the FTW was a company union and that the company was assisting the FTW in its efforts to persuade operators to stay in the federation. Late in the campaign the CWA even obtained a writ to restrain Hazel Magee, Bert Johns and Norm Gravel from interfering in the traffic division. The FTW countered CWA charges by pointing to the large amount of money and staff that the CWA had at its command, suggesting that some traffic officers had been promised positions with the CWA if the raid was successful. The CWA vehemently denied this.[22]

The FTW argued that the CWA was an American-based international union interested in American workers first and

Telephone Talk

In the 1950s BC Tel introduced a portent of the future, an electronic "secretary" that, the company boasted, never needed a coffee break.

foremost. It also pointed to a number of futile attempts by Canadian unionists to break away from internationals; internationals were far easier to join than to leave. The FTW leadership warned members about the loss of autonomy and criticized the CWA constitution and the union's power to revoke charters. They also pointed out that many workers for whom the CWA bargained were not members of the union, and suggested that the CWA should spend its money organizing its own work force instead of raiding organized telephone workers. But above all the FTW leaders concentrated on the bread and butter issues. Could the CWA deliver a better contract for British Columbia telephone workers? While they conceded that the CWA in Saskatchewan had a good contract, this was not the case in the American communications industry as a whole; the CWA had not been able to deliver for its own members in the United States.

The CWA countered by outlining the advantages of belonging to a large international union. Most FTW members accepted that their educational program was weak where it existed at all, and the CWA had demonstrated that it could provide basic union and steward educational workshops. As well, the CWA argued that the operators would become part of the "legitimate" labour movement, at the city, provincial, and national levels. The secretary-treasurer of the BC Federation of Labour, C. Pat O'Neal, attended the first meeting of the CWA Local C-14 and assured members that they would be welcomed into the BC Federation of Labour. The CWA finally pointed out that they employed a permanent staff of experts including research analysts, economists, statisticians, public relations staff, and pension actuaries, all available to member locals.

Near the end of November, after 428 operators had signed with the CWA, Moran and Renaud filed an application for certification of all permanent staff in the traffic division at BC Telephone. The CWA, however, discovered that there were 916 permanent staff as well as 415 temporary and part-time employees. Clearly they did not have a majority of the operators, permanent or otherwise, so they withdrew their application. The raid formally ended in December, 1963. The CWA office in British Columbia was closed a few months later.[23]

In the new year, after the failure of the raid, Elizabeth Lillquist resigned as general secretary at the January council meeting and was replaced by Hazel Magee. A few months later the charter of the New Westminster local was revoked when members, still loyal to the CWA, refused to attend local meetings called by the new general secretary. The proposed merger between the traffic and clerical divisions was rejected by an overwhelming majority of traffic locals. In the fall of 1964 Doug Booth replaced Norm Gravel as general secretary of the clerical division. In just over one year all three general secretaries had been replaced and a new leadership had taken over.[24]

The CWA made a number of errors in its attempted raid. Probably the most obvious and costly error was to concentrate at the beginning on operators outside of the Lower Mainland. While the CWA organizers were travelling the province visiting locals of a few dozen operators, FTW supporters were lining up operators in Vancouver. Furthermore the CWA was taking a gamble in signing up only the traffic workers. As the Canada Labour Relations Board had granted a single certification for plant, traffic, and clerical in 1949, it was likely that the board would deny the traffic division's request for a separate certification, even if they had a majority. The CWA also probably underestimated the resolve of general secretaries Gravel and Johns to fight the campaign. Bert Johns in particular was a long-time opponent of affiliation with any international. Coming from Victoria — the local most vehemently opposed to the federation officers' support of the CWA a decade before — Johns was an old hand at arguing the case against the CWA and in favor of the FTW. Johns' election as general secretary a few months before the raid also helped to "convince people that there would be change."[25]

While there was no proof of direct intervention by the company, the CWA complained that chief operators and their assistants had "suddenly become experts on what is wrong with the CWA." In light of the company's interference during the vote on CWA affiliation a decade before, it is likely that supervisors were again encouraged to expound the merits of an "independent" union over an international union.[26]

The FTW was insulted and angered by the raid. While the leaders had accepted the CWA's courtship, they had not expected a raid. They assumed that the final decision whether to join one of the internationals would be made by referendum ballot of the total membership of the federation. This had been the case on previous occasions. But the CWA wanted to avoid such ballots. As the FTW constitution required a two-thirds majority in order to affiliate, the CWA probably felt that it stood a better chance at succeeding through a raid, since a representation vote would be decided by a simple majority. The raid had been a costly affair for the FTW, though not nearly so costly as it was for the CWA. Joseph Beirne, president of the CWA, was quoted a few years later as saying that the CWA had spent $500,000 in ten years of trying to organize the workers at BC Telephone.[27]

Six months after the raid, the FTW received a letter from the CLC informing the union that its constitution had been amended and that affiliation by the FTW was now possible. The IBEW, which had opposed the application of the FTW in the past, informed the FTW that it would not oppose a new application. At the March, 1965, meeting of the CLC executive council the FTW was admitted into the congress, effective April, 1965. George Meany, head of the AFL-CIO in the United States, attacked the admission of the FTW. In a press conference Meany repeated the CWA charge that the FTW was a "company union," but the CLC refused to back down.[28]

While the union was racked with internal dissent, raids, and reorganization, the company was going through its own transformation. In October, 1955, Theodore Gary and Company merged into General Telephone, with General obtaining controlling interests in BC Telephone in 1956. In 1959 Sylvania Electronic Products Inc. merged with General Telephone to form the General Telephone and Electronics Corporation (GTE), a multinational with interests in Canada, Mexico, Japan, Argentina, Brazil, Italy, Colombia, Panama, the Philippines, Venezuela and the West Indies, as well as 26 telephone companies in the United States.[29]

Although BC Telephone had been owned by an American company long before this latest takeover, a combination of a resurgence of Canadian nationalism in the sixties and the retire-

A lineman, a member of the plant division.

Telephone Talk

ment of Gordon Farrell drew attention to the fact that BC Telephone was owned by an American multinational. Gordon Farrell, son of a founder of the company, had been president since 1928. Most people believed that the company was still a family-owned enterprise. After the takeover Gordon Farrell assured the public that no changes in personnel or policies were contemplated. Within two years of the takeover, however, Farrell stepped down as president and was replaced by Cyrus McLean.

McLean had worked for various Gary subsidiaries, most recently North-West Telephone. But he had started in the industry at BC Telephone and was viewed as a "local man."

McLean instituted few changes and represented a continuation of the old regime. The new broom, however, came in the form of McLean's replacement, J. Ernest Richardson, who took over as president in 1963. Richardson instituted changes which contributed to the workers' growing militancy. As Bert Johns recollects,

"Richardson brought us to the point where we recognized that we were working for a foreign-controlled company."[30]

From 1959 through the sixties not a single contract between the company and the union was concluded without the intervention of federal government conciliation services. Industrial relations at BC Telephone began to break down. Before 1959 there had been only one instance of conciliation in over 30 years of bargaining, in 1951. Lillquist speculated that with the takeover, BC Telephone management no longer wanted responsibility for a settlement and preferred to have all negotiations settled through conciliation. The bargaining committees suspected that the company was no longer bargaining in good faith.[31]

During the 1959 negotiations the union came close to taking strike action. The powerful plant division ratified the agreement by only 52 per cent, while traffic and clerical accepted the agreement by 93 per cent and 83 per cent respectively. In the following year's negotiations the ratification votes were similar. Federal Department of Labour conciliation officer Don Tysoe warned Pettigrew that FTW was heading for trouble, with the possibility that two of the three divisions might accept a settlement while the third rejected it. What in fact happened was that in each succeeding set of negotiations, the traffic and clerical divisions' support for ratification decreased until, by 1969, all three divisions overwhelmingly rejected the conciliation offer.[32]

In the sixties the image of the company as providing "a good job, and a good place to work" was undermined. Wages did not keep up with inflation; pensions, vacation rights, and services which BC Telephone had established decades before were now offered either by federal government programs or by other large employers. "Nobody ever quit BC Telephone before 1955," explained Johns. "Up until that time you didn't quit because the story went around that you had such a wonderful pension when you retired and the conditions were so good." After that year, however, the low starting wages and threat of layoffs fired the militancy of the younger workers. For the older workers, changes in the organization of work and the attitude of management generated unprecedented resentment.

One of the most provocative actions taken by Richardson early in his presidency was the issuing of a "BC Telephone Code of Ethics." This six-page booklet, issued to every worker, outlined the "personal integrity" and "moral obligations" expected from each employee. The code stated:

> Personal integrity should be the prized possession of every individual and safeguarded as a valuable asset. Its maintenance requires vigilance to ensure that honesty prevails under all circum-

> stances. The employee who give a day's work for a day's pay, restricts personal phone calls and holds absences to a minimum is demonstrating his personal integrity.[33]

The booklet stressed that employees should not steal and that secrecy and discretion were required in the communications industry. The patronizing tone and the arrogance exhibited by the company in issuing such a manifesto infuriated the older workers in particular. Most felt that the company was not in a position to preach "moral obligation." The code was seen as an insult.

In issuing the "Code of Ethics" the company was serving notice on the employees that long-established customs such as workers taking extra telephones home and installing free extensions in their homes would no longer be tolerated. But the heavy-handed and self-righteous manner in which the company handled the matter only increased antagonism between management and workers.

The introduction of a wave of efficiency experts in the clerical and traffic departments, subjecting the workers to increasing management scrutiny, further eroded the working relationship. Soon plant division installers were being warned against having coffee together, as groups of trucks parked outside coffee shops was bad for company public relations. Each of these changes might seem insignificant in isolation, but together they created increasing annoyance and dissatisfaction with management.

Older workers felt that the changes marked the end of the old honor system. "We had an honor system in the old days," describes Art Marsden. "We'd try and do eight installations a day and we didn't have to call in — we'd just go out and do our work." But the new system required workers to constantly call the dispatch office so supervisors could keep tabs on where they were. New regulations were drawn up. Analysts were added to each department to make recommendations on further changes to the organization of work. Most workers traced the decline in job satisfaction, increased supervision, reduced job security, and ongoing bargaining problems to the changes instituted by the new owner, GTE.[34]

By the mid-sixties union officers began to speak more frequently about possible strike action in order to win union security. The union had changed its bargaining demand from the closed shop, where all employees were required to be union members, to the Rand formula. The Rand formula is a compromise first established by Justice Ivan Rand in an arbitration award arising out of the Ford Windsor strike of 1945. It gives the union a dues checkoff for all employees in the bargaining unit, in recognition of the services and conditions provided for all employees by the union. Unlike the

Telephone Talk

Telephone Talk

In the 1950s BC Tel computerized its files. Punching IBM cards became a clerk's main function.

closed shop, employees under the Rand formula do not have to join the union. The telephone company steadfastly refused to grant the union either the Rand formula or the closed shop until 1966, when it appeared that the union was willing to strike over the issue. The Rand formula was granted, along with a wage increase of 16.5 per cent in clerical, 19 per cent in plant, and 27 per cent in traffic. The new militancy, coming as it did in a period of record-breaking profits at BC Telephone won the day for the union.[35]

The development of militancy among telephone workers did not take place in isolation from the rest of the labour movement. By the mid-sixties the Canadian labour movement was in the midst of one of the largest strike waves in its history. Unlike the major strike waves which followed the two world wars, this upsurge followed the prosperous fifties. Few predicted a militant outburst, but the recession in the late fifties and early sixties had eroded earlier achievements of organized labour. Unemployment increased, wages did not keep up with inflation, and a new, tough stance by management along with growing restrictions on unions fuelled the discontent. Technological change and automation also had become a concern for many workers, with the issue at the centre of a number of important strikes that hit, among others, the oil refineries, printers and railways.[36]

In British Columbia one of the most significant conflicts broke out shortly after the FTW affiliated to the BC Federation of Labour — the strike of the Oil, Chemical and Atomic Workers' Union at the province's oil refineries. A central issue was the demand for notice of any proposed technological change and protection against job loss, demotions, or pay reductions through automation. Because a significant degree of automation already existed in the industry, management was able to maintain operations by using supervisors and executives as strikebreakers. Management also resorted to injunctions. These were court orders banning pickets. The BC Federation of Labour, recognizing that the use of injunctions was undermining the workers' right to strike, called for a 48-hour general strike.[37]

The FTW initially voted to participate in this sympathetic strike. But a number of unions — for example, the Teamsters and the Machinists — decided on a less militant sympathy action; they opted for a boycott of "hot" products — products produced by strikebreakers. The FTW decided to support the boycott instead. Although the strike did not take place, the threat of a general strike forced the intervention of the provincial government. A settlement was reached with a labour-management committee assigned to draft an automation agreement as part of the contract.[38]

A year after the oil workers' strike came the violent Lenkurt strike. While Lenkurt Automatic Electric was not part of BC Telephone at this time, the plant was nevertheless owned by BC Telephone's parent, the multinational GTE. Management's attempt to make workers at the Lenkurt plant work overtime during contract negotiations sparked a wildcat walkout. An injunction against picketing was served. The workers defied the injunction and a demonstration was called at the Lenkurt site. Police arrived and arrested dozens of workers who were charged with contempt of court for defying an injunction. While 22 pickets were given minor fines and jail sentences, four labour leaders were singled out for jail sentences of three to six months.

Lenkurt sparked an all-out campaign by the labour movement against the use of injunctions, the jailing of labour leaders, and restrictive labour legislation imposed by the Social Credit government of W. A. C. Bennett. The battle against injunctions and the province's anti-labour legislation did not directly affect the majority of FTW members, who fell under the jurisdiction of the Canada Labour Code. But the campaign gave the FTW valuable experience. At stake were such basic labour rights as the right to picket, the right to strike, and labour unity. For nearly 40 years the telephone workers had been isolated from the organized labour movement in the province. Through the labour campaigns of the late sixties the union once again became part of the broad labour movement.[39]

In a few short years the FTW had gone through a major

transformation. It had moved from a union accused by its own members and members of other unions of being a company union, to an active and recognized union participating within the militant BC Federation of Labour and Canadian Labour Congress. Yet for some the FTW's legitimacy within the labour movement would not be secured until the union decisively stood up to the company and backed its demands with strike action.

In 1969, to the utter surprise of BC Telephone, events had come a full circle. After 50 years the telephone workers of British Columbia once again backed their demands with strike action.

Chapter 7 *First Strike in Fifty Years*

The 1968 contract negotiations followed a pattern developed over a decade of bargaining between the Federation of Telephone Workers and the BC Telephone Company. There would be a long-drawn-out period of bargaining during which few items were resolved. Then, as the expiry date of the contract approached, either the union, the company or both would request the appointment of a federal conciliation officer. With the assistance of this third party there would be limited progress in negotiations. Finally, four to five months after the expiry date of the previous contract, a new contract would be signed. Between contracts the company was reluctant to meet with the union for discussion or consultation. Consequently many of the issues which ended up on the bargaining table would in normal industrial relations be resolved by discussion, mutual agreement, or a letter of understanding during the life of an agreement.[1]

Bargaining for a new agreement began in September, 1968. The contract was due to expire at the end of December. By late December there was little progress in negotiations. The union asked for a federal conciliation officer and Donald Tysoe was assigned to the dispute. Tysoe was an old hand at BC Telephone-FTW disputes. He had been conciliation officer in three sets of negotiations between the company and the union. At the outset Tysoe noted that there was "more militancy this time than in the past." Before the conciliator entered the dispute, the plant division had banned overtime and started a work-to-rule campaign. At Tysoe's request the plant general secretary, Bert Johns, agreed to suspend the campaign during mediation.[2]

The central issue was money. The company offered a 6 per cent wage increase in each year of a two-year agreement. The union demanded a 25 per cent increase in one year. The craft workers sought parity with IBEW workers at BC Hydro. Even the traditionally conservative clerical division was prepared to fight on the issue of wages. Doug Booth, general secretary of the clerical division, argued that clerical workers needed a 25 per cent increase in one year to raise clerical wages to a level where a member could "maintain a proper standard of living." Booth pointed out that negotiated increases in some trades had "exceeded the total take-home pay of lower classified clerical employees."[3]

While the dispute centred on wages, long-standing grievances were at the root of the new militancy. The negotiations in 1966 might have ended in a strike if the company had not granted the union's demands for a substantial wage increase and the Rand formula. Many workers thought that the gains won in 1966 had merely postponed the inevitable. There was a widespread feeling that a strike was necessary to change the company's attitude toward the union and the workers. The union's position was that only a strike could establish its authority in the workplace. It hoped that militant action would force the company to consult with the union between contracts and acknowledge that the union was a body which could not be ignored. In the words of Bert Johns, "Richardson thought he had the divine right of kings and you had to be passive and obedient to whatever he decreed." As the 1968 negotiations dragged on into the new year, a growing number of union members resolved to take strike action.[4]

By spring, mediation had failed and a conciliation board was appointed. The report issued by the conciliation board in early May unanimously recommended a 17 per cent wage increase over two years and a reduction in worktime for plant workers, from 40 hours per week to 37½ hours. The company accepted the recommendations, but by now the union was out for blood. The union bargaining committee recommended rejection and the membership of the three divisions overwhelmingly rejected the proposed settlement. The vote was 87 per cent for rejection in plant, 84 per cent for rejection in clerical, and 81 per cent for rejection in traffic. The union held a strike vote during the first two weeks in June.[5]

While the strike vote was being conducted, antagonism between the company and the union grew. The previous month Richardson had distributed a confidential letter to all supervisors and management personnel explaining that it was their responsibility to maintain service in a strike. The FTW leadership viewed the letter — leaked to the union at the beginning of June — as a provocation. The union charged that the letter was leaked in order to intimidate employees and influence the strike vote. Gordon MacFarlane, BC Telephone vice-president of operations, denied that the company was using scare tactics. The company, he explained, was merely gearing up in order to "have some 1,300 management people ready to undertake the necessary tasks to keep operating systems for which some 6,600 people normally are employed."

Part of the company's preparation consisted of training plant and clerical supervisors to use operating equipment. This action infuriated operators because it was clear that the company was training supervisors for strikebreaking. In addition the company took steps to increase its internal security. Manhole covers in downtown Vancouver were sealed to bar access to underground wiring. Security guards were hired and assigned to check the identification of people entering company headquarters at 768 Seymour Street. Plant employees were required to turn in their company keys to supervisors at the end of each shift.[6]

Jim Bremner, an assistant plant general secretary at the time, charged that these security efforts were part of an attempt to "make the employees look like they are contemplating sabotage." The demand that keys be turned in daily, along with the company's endless security checks, were viewed as insults. The union ridiculed the sealing of manhole covers, as the tar used to seal the covers was highly flammable and the covers could be reopened for repairs or inspection only by using welding torches.[7]

In response to the union's charges, McFarlane claimed that the additional security was because of a bomb threat. The company also argued that by taking away employees' keys, it was protecting them from suspicion should there be any damage to company property. The union pointed out that the company had not taken the bomb threat seriously, as no one had been evacuated at the time. As a show of good faith, the union offered a reward of $1,000 to anyone giving information leading to the arrest and conviction of any person who wilfully caused damage to company property or made a bomb threat.[8]

As pressure for a strike grew, the company took its case to the public. On June 6, BC Telephone published a full-page advertisement in newspapers throughout the province expressing its view that telephone workers in the province were the highest paid in Canada. While the company "must be fair with its employees," the advertisement reasoned, it must also "protect the interest of its customers, its investors and British Columbia as a whole." The company explained that it had accepted the conciliation board's report and would not accept any further demands. The advertisement concluded by assuring the public that in the event of a strike the company was "prepared to maintain telephone service for our customers."[9]

The plant division responded with a smaller advertisement the following day defending the bargaining demands of the union and explaining the reasons for discontent among BC Telephone workers. "Do you know," asked the union, "that the morale of the employees, since the takeover by GTE and as a result the introduction of their methods, which treat an employee as a machine rather than a human being, has deteriorated to the point that the employee's respect for his employer no longer exists?" The union criticized the incremental steps which required employees to work six to eight years before receiving full rates and attacked the company's refusal to hire more staff as well as its excessive demands for overtime. On the wage issue the union argued that their members' wages had been consistently falling behind the wages of other skilled workers in British Columbia.[10]

On June 10, even before the strike vote had been completed, the FTW held a mass meeting of telephone workers in the Lower Mainland. At the meeting, attended by 3,000 members, the union revealed its strike strategy. The union proposed that in the event of a strike, workers would engage in selective rotating strikes and form travelling pickets called "flying squads." This tactic had been used successfully by Hydro workers in Ontario and Quebec.[11]

Workers striking a highly automated utility such as Hydro or the telephone company need innovative tactics in order to apply economic pressure on their employers. Workers in other industries have the power to stop production and the creation of profit simply by laying down tools. But utility workers have limited economic leverage on their employer. In these highly-automated industries a strike does not automatically stop the delivery of service. Nor does it interrupt the company's collection of revenues. Even though service declines, a relatively small management force working as strikebreakers can keep the network functioning. The company also saves millions of dollars in workers' wages. As monopolies, utilities do not even risk a loss of customers because of declining service. Most unions confronting a monopoly employer try to bring public pressure to bear on the company. But this approach is seldom successful as the public is generally indifferent to strikes while they are still receiving service. Worse still, because management is seen as delivering the service to the public, it is the workers who come under pressure to settle.

DDD—direct distance dialing—automated long distance calling.

Telephone Talk

Flying squads of mobile pickets allow the union to take the strike to the customers of the company. This makes it possible to apply economic pressure. Flying squads also shut down much of the company's construction, installation and repairs on unionized worksites, because union workers in British Columbia will not work behind a picket line. Rotating strikes interrupt service while most of the work force remains on the job, ensuring that the company does not profit from a massively reduced payroll. It also demonstrates to the public the union's willingness to maintain service.

When it heard of the union's proposed strategy, the company warned that it would not tolerate a rotating strike. MacFarlane declared, "When management people are put to work to maintain service in a strike situation, there may be no work for employees who have been on strike." In other words, once management personnel had been put on duty in any area, they would remain on duty for the duration of the strike. The company's tactic was clear: it would counter rotating strikes with lockouts.

Management was confident that it could maintain operations throughout a strike. Company spokesmen informed the conciliation officer that "no telephone union has ever won a strike" even though strikes in the industry "have on the average lasted seventy days." The company felt it could afford to wait out a strike while the workers could not.[12]

By the middle of June the strike ballots were counted. A stunning 86 per cent of the membership endorsed strike action. The union made a final offer of a two-year agreement with a wage increase of 18 per cent in the first year and 12 per cent in the second. This effort failed. On Monday, June 23, rotating strikes began to hit operating services in the Lower Mainland. This was the FTW's first strike and the first strike at BC Telephone since 1919.[13]

For the first few days only the Lower Mainland was struck. But towards the end of the week workers in the interior and on Vancouver Island were becoming frustrated at having to remain at work. Walter Torry, the FTW president, told reporters that members found work "intolerable" because supervisors were constantly "looking over their shoulders to learn their jobs." The company's threat that workers who struck would be out for the duration of the dispute brought the rotating strikes into question. The rotating strike also made it possible for the company to send supervisors from locations not yet struck into areas where there was a bottleneck caused by job action. On June 27 the strategy of limited strike was scrapped. A full-scale strike began.[13]

Although this was the first strike in 50 years at BC Telephone, the company was prepared. Management knew that a skeleton staff could maintain service for a long time. Most supervisors were assigned to operating services, where they worked as directory assistant operators, toll operators, or at the automatic direct distance dialing equipment. A number of plant supervisors were assigned to emergency repairs and maintenance. By working twelve-hour shifts seven days a week, the 1,500 strikebreakers were able to keep the network functioning.[14]

The strike caused little inconvenience to the public. Long distance calling became more difficult. By 1969, most subscribers could dial their own long distance calls, but an operator wrote up a ticket on each call and submitted it to the billing office. Supervisors staffing this system during the strike could not keep up with the call load, which resulted in calls being disconnected or subscribers receiving busy signals before they had finished dialing.

Requests for a new telephone, repairs or inquiries were impossible; only emergencies were handled. But for the majority of subscribers the strike was a novelty in that inexperienced operators were answering calls.[15]

BC Telephone normally hired hundreds of post-secondary

students as temporary workers in the summer. At the end of June many students reported to work in spite of the strike. Hundreds more refused to cross picket lines. Most students who worked as strikebreakers were engineering students who claimed to be management trainees. The BC Union of Students, comprising representatives from university and college student councils throughout the province, issued a press release supporting the FTW's strike action and condemning the company for its use of students as strikebreakers.[16]

Early in July the FTW started to send out flying squads. Mobile pickets travelled to buildings and worksites where supervisors were doing installation and repair work. Within days a BC Hydro plant, a post office in Terrace, and a major office complex in Vancouver were struck. The BC Federation of Labour responded to a request from the FTW and declared that all telephone conduit being installed in commercial and industrial projects were "hot." The construction trades abided by the declaration, bringing the pre-wiring of new buildings and the installation of telephone conduit to a halt. Ed Faye, secretary of the BC and Yukon Building Trades Union Council, warned employers that construction workers "would not work on any sites where there are non-union workers." This solidarity action by the construction trades generated strong economic and political pressure on the company. "Every day this strike is allowed to drag on," warned Ian Edgar, assistant clerical general secretary, "more construction work bogs down because the phone equipment can't be installed."[17]

During the strike, flying squads hit a wide range of businesses and industries. Many of the companies struck by the flying squads challenged the legality of these pickets, but the courts ruled that pickets were legal as long as company crews were on the premises. Wherever the supervisors worked was a "place of operation and a place of employment of the company."[18]

Midway through the strike the FTW attempted to persuade supervisors to join the union in strike action. The effort was aimed primarily at the foremen and first-line supervisors. The union argued that "these supervisors are working people, not highly paid brass." President Walter Torry assured supervisors that the union would protect the jobs of any supervisors who joined the strike. Picket signs and leaflets carried the message asking supervisors to "join their fellow workers rather than support the efforts of the New York-controlled monopoly to break the strike of telephone employees." But this effort to win over the supervisors failed.[19]

The issue of nationalization of BC Telephone re-emerged during the strike. Calls for nationalization of the foreign-owned telephone company was raised on numerous occasions in the thirties by the Co-operative Commonwealth Federation, the predecessor of the NDP and official opposition in the British Columbia legislature. When BC Electric was taken over by the province in the early sixties, the call for nationalization of the telephone company was revived. More than one newspaper editorial asked the Social Credit government of W. A. C. Bennett, "Why stop with BC Electric?" Bennett passed off calls for a government takeover of BC Telephone by arguing that "Alexander Graham Bell didn't want it himself, so why should the government?" In the summer of 1969 there was much speculation that a provincial election would be called. When the election was called at the end of summer, a key issue became public ownership of the telephone company. NDP leader Tom Berger reiterated the position that an NDP government would take over BC Telephone.[20]

By mid-July pressure began to build for an end to the strike. The flying pickets, declining service, and growing discontent of the business community led to calls for the federal minister of labour, Bryce Mackasey, to intercede in the dispute. At first Mackasey rejected such calls, claiming, "A strike is a strike . . . both parties knew what was happening." But as the strike entered its second month, Mackasey instructed conciliation officer Don Tysoe to bring the parties back to the bargaining table. Negotiations began again on July 24.[21]

For five days negotiations yielded little progress. Finally the company agreed to a slight increase over the 17 per cent wage boost recommended by the conciliation report, offering a 19.84 per cent increase over two years. The union was unimpressed. At the insistence of the conciliation officer, however, the bargaining committee agreed to submit this new offer to the membership. When it became clear that the union would not recommend acceptance to its members, MacKasey intervened. In private meetings with the union Mackasey pressured the bargaining team to recommend acceptance of the offer. At 3 am on July 31 a tentative agreement was reached. The telephone workers' first strike in 50 years was effectively over, though the strike would not officially end until the agreement was ratified by the membership.[22]

Only 65 per cent of the membership voted to accept the new agreement. Many members felt that the strike should continue, at least for a few more weeks. Phil Zander, a craft worker, recalls voting against the settlement. He thought the union "was only two weeks away from getting what we wanted." Theresa North, an operator, stated with amazement, "I couldn't believe that people would be willing to go back to work after six weeks on strike, for only a few cents." Others argued that the six-week strike had taken its toll on the workers. The leadership of both the clerical and traffic divisions was concerned that they could not "hold their members." Doug Booth, clerical general secretary, described the settlement as a "disappointment." "We should have got more," explained Booth, "but I didn't see that there was much more without staying out for a very long period of time. Even then, there was no guarantee we would get any more." Bert Johns defended the settlement, recalling that "towards the end, things had started to get rough." Johns argued that "we were starting to lose support for the strike . . . and if we had not settled when we did, we would have destroyed the union."[23]

From the outset the strike had two goals. One was to improve wages and the other was to assert the union's authority. The small 2.84 per cent increase over the original offer was minimal after a six-week strike. Yet few members thought of the strike as a failure. The company had been forced to increase its wage offer, despite its statement at the beginning of the strike that the union could get no more. Most hoped that the union's militant action had demonstrated to the company that the employees' wishes could no longer be ignored.

Most members did not measure the gains won by the union in dollars and cents. The confrontation brought the union together more than any other event in its history. This strike marked the beginning of the "one union movement." For the first time the plant, traffic, and clerical workers worked together at the rank and file level. For six weeks they walked the picket line together; operators, clerical workers, and craft workers got to know each other. Before the strike there was little contact between the divisions. Each division council met separately. At the beginning of the strike each division even had different rules on strike pay, although a common position was eventually worked out. In the course of the strike members gained an appreciation of the problems faced by workers in the other divisions. The strike also engendered a feeling that the union needed to change its structure. Members wanted to bring the divisions closer together in an organization which functioned as a single union.

In spite of its length the strike was a relatively easy one. It took place during the summer; picketers were frequently pictured in the newspapers wearing bathing suits. Strikers in Kamloops held pancake breakfasts on the picket line and for many weeks a festive atmosphere prevailed. The union paid $25 a week strike pay. Though this was hardly a generous sum, it could buy a week's groceries in 1969. Even though the company attempted to sensationalize a few incidents of cable slicing and damage to company property, BC Telephone president Ernest Richardson was forced to compliment the union on "a clean strike." Supreme court judge

Peter Seaton, who heard most of the employers' applications to restrain the union from picketing, praised the FTW's conduct during the strike.[24]

There were a few instances where supervisors taunted picketers by waving their inflated paychecks at strikers, but such incidents were the exception. Dolly Storey, an operator and a picket captain during the strike, recalls the arbitrary distinction often made by picketers between supervisors and "scabs." "Scabs" were students, Rand formula employees, and the few union members who crossed union picket lines. These people were taunted by the strikers. Supervisors, on the other hand, were viewed as doing their job and therefore not harassed. In subsequent strikes, however, members came to realize that the supervisors did not have to cross the picket line and that by crossing they were strikebreaking.[25]

The layoffs in the early sixties first drew the union's attention to the threat posed by automation. But the strike at the end of the decade drove the point home. At the start of the strike many believed that the company could not maintain service without the 6,600 striking workers. Jim Bremner notes that many workers thought that the union would be legislated back to work after a few days. Once on the street, however, telephone workers discovered

Everybody in management was expected to pitch in and help break the 1969 strike. Here, BC Tel president J. Ernest Richardson poses for photographers as an installer's helper. Note that he is holding the test set upside down.

that the combination of a highly-automated industry and a large non-union staff provided BC Telephone with an effective strike-breaking force.[26]

As the telephone workers returned to their jobs in the first week of August, there were heated confrontations. In spite of a clause in the back-to-work agreement that there would be no victimization after the strike, the company was intent on punishing employees who had gone on strike. Jack Wasserman, a Vancouver *Sun* columnist, referred to the company's post-strike behavior as a "dog-in-the-manger attitude," with BC Telephone insisting "that the operators walk over hot coals in order to get to their switchboards." Wasserman felt "that company officials still haven't recovered from the shock of a well-kept, virtual company union getting up the guts to hit the bricks."[27]

The strike had given the unionists an appreciation of their collective strength. They were not prepared to tolerate discrimination or victimization by the company. On the first day back at work there was a wildcat walkout by plant workers at the Welwyn Street construction and installation office. This walkout protested the company's attempt to victimize a student summer employee who had refused to cross FTW picket lines. The student had been called in to work at the end of the strike. When he arrived at the compound, however, supervisors told him that he had been called by mistake and there was no work for him. The walkout was settled within a few hours when the company agreed to find work for the student.[28]

Clerical workers at a number of Vancouver offices also threatened walkouts on their first day back on the job. The company had posted a Tuesday-to-Saturday work schedule. The clerical workers objected to working on Saturday unless they received more pay. The company defended its new schedule. Since the workers had returned to work on a Tuesday, the company was simply giving clerks an opportunity to work a full week. The company backed down under the threat of strike action and no one was required to work on the Saturday.

In the Kootenays the company decided to call all workers back to work at once with the exception of Nelson. Management proposed bringing the Nelson workers back at staggered intervals. After union protests the company agreed to recall all employees at the same time.

The union was less successful in its fight against the disciplining of a supply record supervisor. A woman who became a supervisor shortly before the strike was fired when she refused to cross FTW pickets. The union, working with conciliation officer Tysoe, won the woman's reinstatement. But she was not rehired as a supervisor and returned instead to the position she held before promotion.[29]

The only violent incident occurred several hours after the strike was settled. A few strikers jumped a group of supervisors coming off shift in Nanaimo, hospitalizing one of the supervisors. The company fired three employees accused of participating in the fight. However, only one of the fired employees was convicted in court. One of those fired was Fred Barnes, president of the Nanaimo plant local and a worker with the company for over twenty years. Barnes did not participate in the assault and no charges were brought against him. But the company insisted that as a strike leader Barnes was responsible for creating the hostile atmosphere which led to the assault. Bert Johns contended that company president Richardson "was determined to make an example of somebody for the strike" and Barnes was his scapegoat. The union fought the firing of Barnes and another employee fired in the incident. The other employee was reinstated, but Barnes was not. The union did not give up on the case. More than three years later the union won Barnes his accumulated pension benefits and a job with a Victoria company by instituting a province-wide ban on overtime.[30]

With the exception of the Barnes case, the 1969 strike did not generate a great deal of bitterness between the union and the company. The antagonism which erupted after the return to work eventually dissipated. In the weeks after the strike the company spoke of improving operator and junior employees' jobs. Richardson told a reporter that supervisors staffing the boards during the strike had discovered the jobs to be boring and dull. "We're going to look at all the repetitive jobs and make them more interesting, less frustrating and give the junior employees a broader experience," promised Richardson. The union, he assured, would be fully involved in this work reorganization. But no action was taken to improve "the dull jobs," nor was the union invited to work with the company on such a job study. The proposed restructuring of jobs was an instance where the company could have responded to the concerns of its employees and opened a dialogue between the union and the company. However, the matter never went any further. After the initial flurry of company publicity, the plan to restructure jobs was dropped.[31]

On their return, the employees discovered a massive backlog of work. The situation was especially critical in the plant division, where many subscribers were waiting for installations and repairs. The company promised the public that the backlog would be cleared within weeks. At the same time, however, the company refused to authorize overtime by installers. Finally, with the company facing widespread criticism for its failure to provide adequate service long after the strike was over, BC Telephone reinstated overtime in October.[32]

Soon after the strike, a further attempt to launch a national union of telephone workers began. In late August the FTW received a letter from Boris Mather, the executive director of the Canadian Communications Workers' Council (CCWC). The council comprised the Canadian locals of the CWA, which were in the process of leaving the CWA and forming a separate Canadian union. Mather suggested that merger discussions take place. The proposal came at an auspicious time for both the CCWC and the FTW. The Canadian locals were hesitant to take the step to full autonomy. They feared that such a move would reduce their strength and jeopardize access to the strike fund available through the international union. Its recent strike left the FTW receptive to suggestions which could increase its bargaining strength. Mather assured the FTW leaders that while a merger would require the CWA locals to leave their international, the Canadian locals were already in the process of leaving and "there would be no opposition from the international union to such a move."[33]

Discussions between the CCWC and the FTW continued for a number of years. By 1972 a draft constitution for the Canadian Federation of Communications Workers (CFCW) was ready and in the spring it was ratified by the three divisions of the FTW. The constitution stressed that membership in the CFCW was confined to Canadian unions. It guaranteed the autonomy of member affiliates in bargaining and set up a defence fund with affiliates paying 50 cents per member a month into the fund.

A year after the Canadian locals of the CWA had held their founding convention of a new communications union—the Communications Workers of Canada (CWC)—the FTW and CWC met in Vancouver and founded the Canadian Federationof Communications Workers. Mather was elected CFCW chairperson and authorized to set up an office for the organization in Toronto. More than 25 years after the first discussions on a national union of telephone workers, "the first concrete action" brought together telephone workers from British Columbia, Saskatchewan, and a small group of equipment manufacturing workers from Ontario.[34]

Two years later some craft and service employees at Bell in Ontario and Quebec decided to break with their company union and join the CWC. The CFCW endorsed the Bell campaign. Mather spent most of the next year and a half assisting the CWC campaign. The FTW sent plant business agent Tom Stanley and clerical business agent Chris Andersen to help with the campaign shortly before the representation vote. The campaign was a success and saw over 12,000 telephone workers join the CWC and consequently the CFCW.

While discussions on a national body were taking place, the FTW entered negotiations with BC Telephone. The 1970 negotiations was the only time in the last twenty years that the union and company signed a contract without the intervention of a conciliator. The plant division was represented by its new general secretary, Bill Clark, as Bert Johns had retired in the spring. Clark started with BC Telephone in 1951 as a groundman, and worked his way through craft apprenticeship to lineman. A member of New Westminster Local 7, Clark was elected shop steward and had represented the local on the plant council since the early sixties. In 1968 he became a full-time staffer for the plant division on election as assistant general secretary. During the 1969 strike Clark was in charge of the union's flying squads and had won a reputation within the union as a militant.[35]

In the year following the 1969 strike, the union had noted a "marked improvement" in dealings with the company's personnel and industrial relations department. For the first time in more than a decade, the union found that the machinery for resolving grievances and disagreements over the contract was working. At least part of the improved atmosphere was attributed to the company's new director of personnel and industrial relations, J. H. Bruce. At the end of the 1969 strike the former director, D. M. Carter, was transferred to Florida. Bruce, his replacement, had extensive experience in the industrial relations field. Unlike his predecessors, Bruce was not from the industry and therefore not trained in the paternalist style typical of the telephone company. He had worked with strong and respected unions in the auto industry and the food processing field. The union responded positively to Bruce's approach.[36]

Negotiations began in the fall and continued until the end of the year. In January, 1971, the union ratified a new two-year agreement covering wage increases, improvements in the holiday, vacation, and special leave provisions, and protection against contracting out. The union had been concerned about the company's pension plan. The agreement included for the first time a short statement on the pension plan. The union was serving notice that pensions would be the major issue in the next set of negotiations.[37]

Since the early sixties, when inflation began to erode pension benefits, the union's attempts to make pensions an issue in bargaining had been rejected by the company. Management regarded the pension plan as solely the concern of the company; the union had no voice in the running of the plan. Although employees made no contributions, the union felt that the plan was an employee benefit and that employees had every right to participate in the running of the plan.

From the beginning, talks on the new contract for 1973-74 bogged down on the issues of pensions and wage increases. The company stuck to its position that pensions were non-negotiable. On the wage issue the company was preparing a submission to the Canadian Transport Commission—the federal body which regulated BC Telephone at the time—to request a rate increase. The company wanted to be in the position to defend wage increases on the grounds that they were imposed by a conciliation board. The company reverted to its previous practice of using a conciliation officer to whittle the union's position down to the bare bones, with a conciliation board setting the final offer. So after close to three months of bargaining, a conciliation officer was called into the contract talks. But the conciliator made little headway; his recommendations were rejected out of hand by the company. Early in the new year a conciliation board was appointed.[38]

The slow negotiations and the indications that the company was dragging out the process angered the workers. With a breakdown in negotiations the operators started a work to rule campaign in early January. This return to trench warfare did not sit well with Bruce. After the company rejected the conciliation officer's report, Bruce resigned. Bruce was the first and last director of industrial relations from outside the company. He was replaced within days by Frank Tucker, a 26-year veteran at BC Telephone. Tucker had

On the picket line during the five-month OK Tel strike.

come up through the ranks at BC Telephone and his appointment as head of industrial relations marked the return by the company to its previous style of paternalist management. The union, however, had long since rejected this approach.[39]

In February, lawyer Hugh Ladner was appointed to chair the conciliation board. The central unresolved issues were wages and pensions. The board delivered majority and minority recommendations. Ladner and the union appointee submitted the majority report recommending a 17-per cent general wage increase over two years plus a further increase of 20 cents an hour for operators. The report also recommended an increase in the company's contribution to the pension plan, representation for the union on the pension benefits committee, and the setting up of a supplementary employee-funded pension plan jointly administered by the company and union. The minority report differed primarily on the pension issue. The company appointee was adamant that the union should have no say in the administration of the pension plan.[40]

The union submitted the Ladner report to the membership. Over 90 per cent of the membership voted to accept its recommendations, but the company balked at accepting employee participation on the benefits committee. Finally, in early June, direct negotiations between the company and the union reached a settlement based on the Ladner report.[41]

A further problem developed when the company presented a final draft for publication to the clerical general secretary, Doug Booth. Booth discovered that the company had unilaterally altered a number of classifications in the clerical appendix. Booth refused to sign the agreement, and the company refused to make corrections. When the collective agreement was published, it came out in two separate books. One version, jointly published with the company, included the master agreement and the plant and traffic appendices. The clerical division published its own version of the agreement. Booth never signed the agreement, and the issue of clerical classifications was resolved in the subsequent set of negotiations.[42]

Within weeks of signing the agreement, negotiations began between the FTW and the Okanagan Telephone Company. OK Telephone had become a wholly-owned subsidiary of BC Telephone. As a provincially-chartered company, the workers at OK Telephone were certified by the provincial Ministry of Labour, with the FTW holding three separate certifications for the plant, traffic and clerical locals.

The workers' key demands related to two issues — pensions and wage parity with BC Telephone workers. In the late sixties workers at OK Telephone had won wage parity with their BC Telephone counterparts. But unlike the BC Telephone employees, who had a 100-per-cent-company-funded pension plan, the OK Telephone employees contributed 4 per cent of their wages into their pension fund. The company intimated that it was paying matching funds, but this was not the case. In 1971 the union negotiated an end to employee contributions. But as the company warned that the pension plan was in jeopardy, the plant and clerical employees agreed to accept 45 cents a day less than their counterparts at BC Telephone on the understanding that the difference would be placed into the pension fund. Later the union discovered that none of this money had been put into the pension fund and that the fund actually had a surplus of almost one quarter of a million dollars at the time that the company was crying poverty.[43]

It was difficult for the union to get information about the pensions because the plan was administered solely by the company. When the union attempted to establish who was on the pension committee in addition to OK Telephone general manager D. W. Champion, even the company was not sure. The committee whose job it was to oversee the pension had never even held a meeting. By the summer the telephone workers were prepared to strike on the pension issue alone. The demand for joint trusteeship and an equal voice on the pension board became the focus of the 1973 dispute.[44]

When contract talks broke off at the end of July, the FTW took a strike vote among the 500 workers at OK Telephone. Before the strike vote could be counted, workers walked off the job at Kelowna when supervisors attempted to work alongside bargaining unit employees. The company branded the walkout illegal and the 115 workers at Kelowna were given three-day suspensions when they returned to work the following day. This triggered a new one-day walkout, with workers from Penticton, Salmon Arm, Revelstoke and Vernon joining their co-workers.[45]

With 86 per cent of the members voting in favor of strike action, the union issued 72-hour strike notice to the company. A last ditch effort by the provincial mediator to seek a settlement failed. So at 1 pm on Tuesday, August 21, the strike began at OK Telephone.

It was clear that the company would rely heavily on aid from BC Telephone. Champion had told the press before the strike that OK Telephone "will undoubtedly ask the parent company, BC Tel, for supervisory personnel to assist in keeping phone services going." At BC Telephone the FTW placed a ban on overtime work and relieving of supervisors as such work would only free supervisors to relieve the strikebreakers at OK Telephone. On August 22, BC Telephone responded by suspending 32 workers for refusing to do overtime work. Although the contract stated that overtime was to be kept to a minimum with the exception of emergencies brought about by weather or the breakdown of emergency equipment, the company insisted that workers were required to give adequate reasons for refusing overtime.[46]

The day after the suspensions 4,000 telephone workers in the Lower Mainland went on strike to protest the company's attempt to establish compulsory overtime. The following day the company obtained an injunction which ruled the strike illegal and ordered an end to picketing. But the union had established its point, and overtime remained voluntary. Union members did no overtime for the remainder of the OK Telephone workers' strike.

Working under tremendous stress, plant general secretary Bill Clark suffered a heart attack on the day that the company was granted its injunction. The plant division held a special meeting one week later and elected an assistant general secretary, Bob Donnelly, to the post until Clark was able to resume his duties.[47]

In September and October strikers from OK Telephone stepped up the pressure by picketing BC Telephone offices around the province. In mid-October OK Telephone was granted a further injunction to stop the roving pickets. The union turned to the public and ran a number of advertisements in local newspapers. The union also distributed information leaflets in Victoria and sent a delegation to lobby provincial cabinet ministers. The provincial NDP government, headed by Dave Barrett, had proposed legislation to legalize secondary picketing, and the union urged the government to adopt the proposed legislation as it would have helped to shorten the strike. However, the legislation was not adopted until the following year.

"Is it too much to ask for a decent pension?" asked the union advertisements and leaflets. With almost 1.2 million dollars in the pension fund, the fund had paid out a total of $29,000 to 25 retirees including management pensioners. With sole control of the pensions, management had granted executive officers at OK Telephone and BC Telephone past service credits at whim. One of the most blatant examples of this practice involved BC Telephone president Richardson. As president, Richardson was also president of OK Telephone. Although he was hired at BC Telephone in 1963, Richardson was granted service credits with the Okanagan pension plan back to 1946. This action can be compared to the plight of a telephone operator who had worked 30 years with the company and on retirement at 60 was given a pension of $156.17 a month, which was reduced to $3.12 monthly at age 65.[48]

The OK Telephone workers remained on strike for over twenty weeks. As with the 1969 BC Telephone strike, supervisors and management personnel staffed the boards and maintained basic operating services. Over 80 supervisors from BC Telephone were sent to the Okanagan to relieve strikebreakers. As well, Okanagan supervisors were periodically transferred to BC Telephone for rest and relief.[49]

During the 1969 strike OK Telephone workers had donated 10 per cent of their wages to the FTW strike fund. This solidarity was reciprocated in the form of financial assistance from BC Telephone workers, a grant from the plant benevolent society, and donations from labour councils, the CFCW, and other unions. The BC Federation of Labour declared all poles carrying both hydro and telephones lines "hot."[50]

In November the provincial Ministry of Labour set up an industrial inquiry commission to seek an end to the strike. The commission, which did not meet until late December, brought down a unanimous recommendation. The company immediately accepted the proposed settlement, and the union members voted overwhelmingly to accept the proposal a few days later at a mass meeting in Vernon. The back-to-work agreement included a provision that work would not resume until January 2. The company was anxious to get the employees back to work and relieve the supervisors who had been staffing the switchboards

Supervisors strikebreaking during the 1969 dispute.

since August. The strikers decided to let the supervisors work while they enjoyed the New Year holiday.

The settlement "broke the back of the company on the pensions issue," according to Donnelly. A jointly-trusteed pension was established. The company paid 5 per cent of payroll into the fund and employees contributed 2 per cent of their wages. The Okanagan workers also won parity with BC Telephone workers' wages.[51]

The OK Telephone strike demonstrated an important tactic—the selective strike. One segment of the union had stayed on strike for 133 days and had won concessions later extended to the whole union. The jointly-trusteed pension plan became the basis for the bargaining unit employees' pension plan, which was established at BC Telephone in the next set of negotiations. With 8,000 telephone workers still working at BC Telephone, the union could easily maintain the strike fund for the 500 Okanagan strikers. Strike pay was even increased to $75 per week when it became clear that the strike would continue for months.

Probably no other single issue more clearly marked the end of BC Telephone's paternal reign than the dispute over pensions. The workers had investigated the company's claim that it was taking care of its employees and had found the level of care unacceptable. The fact that the union was forced to resort to strike action to gain a voice on a pension plan which had been 100 per cent employee-funded was revealing. It indicated that the company would go to great lengths to prevent employee participation in administrative decision making, even when the decisions significantly affected only the employees. The 1973 strike delineated the basis of future confrontations: the rights of management versus the rights of workers.

Chapter 8 *Technological Change*

The seventies heralded the computer age at BC Telephone. This massive wave of technological advance was part of a second industrial revolution which continues today. Fueling this transition were increasingly sophisticated electronic equipment and growing applications for computers. The decade saw the beginnings of the convergence of telecommunications, computers and office equipment technologies. The telephone workers grappled with radical changes in tools and materials, increases in productivity, and modifications in work methods and organization.

The telephone industry was among the first to seek to apply the new microelectronics technology. In fact telephone industry research and development played a leading role in developing the solid state semiconductor technology. Bell laboratories in the United States developed the transistor in 1948, thereby laying the basis for the semiconductor revolution. Transistors—first used in hearing aids and telephone switches—by the end of the fifties were found in radios, television cameras and computers. Transistors were faster, cheaper, smaller, more energy-efficient and far more reliable than the earlier equipment based on tubes or valves. Technology continued to evolve rapidly. By the sixties individual transistors were replaced by integrated circuits which included transistors, resistors and capacitors as well as their connections, all printed on a tiny chip of silicon. These chips were complete electrical circuits produced on a microscopic scale. By the early seventies the microprocessor—an integrated circuit with the properties of a complete central processing unit or "brain" of a computer—had been developed. The microprocessor amounted to a computer-on-a-chip. Midway through the decade, tiny microprocessors selling for a few dollars had greater computing capacity than the first computers, which took up large rooms and cost millions of dollars.[1]

Microprocessors rapidly replaced electromechanical equipment. The new generation of "smart" machines represented a qualitative leap ahead; the new machines had memory and computing capacity. The computerized long distance switchboard, the Traffic Service Position System (TSPS), for example, eliminated much of the work of long distance operators. On the old cord boards operators had prepared tickets with billing information and manually forwarded calls. With TSPS the automated system forwarded the call and recorded relevant information, including where the call had originated, where it was going, and how long it lasted. As well, the data recorded by the TSPS could be transferred to the company's billing computer which made up bills directly.

The microelectronics revolution marked the convergence of different computing systems. Computer systems linked to electronic office equipment gave rise to the automated or integrated office. The telephone industry was affected both internally and externally by this convergence. Externally it created a large, profitable market for interactive telecommunications systems which used telephone lines to link the new office equipment with computers. Data transmission over telephone lines became a lucrative and significant sector within the communications industry. At the same time the convergence allowed office equipment producers and computer manufacturers to enter the telecommunications field. The company's end-to-end monopoly on telephone equipment was challenged. At the end of the decade it was broken.

BC Telephone began to use microelectronic equipment to computerize most aspects of its record keeping. A decade before the microelectronics explosion of the seventies, the company had introduced computer technology in the accounting functions; further records were maintained on computer systems, with print-outs and microfiche produced at regular intervals. As the cost of computer technology continued to drop, individual terminals were brought into the clerical area as part of a general proliferation of computerized record storage and updating. Four computer data bases were introduced: the Customer Record Information System (CRIS), which records billing information; Mechanized Assignment Inventory System (MAIS), which provides information on available facilities, for example, cable pairs, terminal number, and telephone numbers not yet assigned; Service Order Update and Locate (SOUL), which keeps track of the service orders; and Left In Facilities and Equipment (LIFE), which records equipment and facilities that are available. Clerical workers using computer terminals for record keeping now fell under a new regime where the system program controlled the operator, creating a rigid and often frustrating dialogue between operator and machine in which the machine had the final say.

In the clerical areas where computers were not introduced, significant organizational changes took place. New methods—modelled on the rigid yes/no flow chart logic of computers—were established. Clerical workers had to use "decision trees," which outlined steps in decision-making and paved the way for the eventual transfer of even the most limited decision-making to the computer system. The rationale underlying mechanical decision-making is that complicated phenomena can be handled by answers to a series of simple questions. John Johnson, a sixteen-year veteran with BC Telephone, indicated some problems with the new "decision tree" trouble reporting system and explained how this system actually created longer waiting periods for customers with service difficulties:

> When I was first in I and R, customers would phone their trouble reports into a repair service clerk who would write the trouble ticket up, pull the proper line card with all the relevant information on the telephone number in question, and then pass it directly to the test desk man. The test desk man is a fully qualified journeyman, trained in testing telephone facilities. The test desk man would test the trouble and dispatch it to the proper area. He would decide whether the trouble had to be dealt with from the central office or dispatch it to the repairmen responsible for outside plant. . . Today, a customer reporting a trouble still phones a repair service clerk. But the clerk now uses something known as a decision tree to record the information given by the customer. The major problem with this decision tree approach is that if the customers do not answer the questions put to them properly, their troubles could be improperly dispatched to central office, for example, when they should have gone to outside plant desk. Even worse, they could be placed into the 3-day or 5-day follow-up bins. . . This has the effect of increasing the amount of time that customers must wait for their phones to be repaired while the trouble is pursued by the wrong people and then while the trouble is re-routed to the proper repair area. The use of the decision trees reduces the need to employ test desk journeymen.[2]

In the central office, computer-controlled electronic switching was introduced when Automatic Electric, the manufacturing wing of GTE, began to produce electronic switching equipment in the seventies. Its electronic exchange, the EAX, required the attention of few craft workers. It did not need the same routine of maintenance as the old electromechanical step-by-step equipment. Much of the central office craft job of troubleshooting, switch repairs and maintenance became a matter of "card pulling." The diagnoses of troubles now amounted to identifying and replacing individual cards containing the malfunctioning integrated circuits.

The traffic division, which was already heavily automated, received a further blow with the introduction of computerized switchboards and computerized directory assistance. The new computer systems automatically routed calls to operators. A call came on the line immediately after the last call was completed. The new computer-controlled operator systems removed what little control the operator still had over her pace of work, and could monitor operators to a fraction of a second and the individual keystroke.

The introduction of computers was another step in the unending move towards total automation of the traffic division. Job satisfaction sank to an all-time low. Linda Rolufs described her experience in switching from a cord board to the automated TSPS board:

> When I began working at BC Tel, I enjoyed my job and took pride in my work. Operators were given the flexibility to handle customers' problems when they arose. The emphasis was on meeting customers' needs. . . When I transferred from the cord board office in Prince George to the TSPS office in Vancouver, I saw a big difference in the way calls are handled. On the cord board an Operator decided when she would plug into an in-coming call. With TSPS the computer connects the Operator to a new call as soon as her other customer is off the line. If your desire is to process as many calls as possible in the shortest possible time, then I guess you could say that this is an increase in productivity. But the increased pace, combined with the fact that the machine is now in control, means that what used to be a human work environment has become like an assembly line. Now drudgery and stress dominate the Operator's work life. . .Operators are constantly pushed to speed up and to handle calls as fast as possible. When you are monitored or evaluated, you are admonished to keep customer contact to a minimum. We actually had one Operator called down for being too nice to the customers.[3]

Integral to the technological changes at BC Telephone was a general reorganization aimed at centralized record-keeping, elimination of operating centres in all but five locations around the province, and a reduction in customer service to outlying areas. Rod Hiebert, a craft worker in central office engineering and an employee of BC Telephone for over seventeen years, outlined some problems inherent in the centralization which accompanied the introduction of electronic switching offices:

> The new technology which BC Tel is now installing puts all the Company's eggs in one basket. In these Stored Program Control offices, certain switching functions are directed by a common on-line central processing unit or computer. If one integral part of the computer fails, if a program fault exists, or if there is a major cable outage, entire switching offices can go completely out of service for hours. This disconnects entire cities from all phone service. For example, the Dilworth telephone exchange, now serving approximately 24,000 customers in Kelowna, has had numerous total outages ranging in duration up to twenty hours at a time. In addition to the disruption suffered by the local phone users, long distance dialing was almost completely severed for an additional 24,000 customers in office. . . In all my time at BC Tel I have never heard of an entire step-by-step office suffering total failure. But office-wide failures in the new EAX offices do occur, and these failures are not rare.[4]

New methods of work were as significant as the new machinery. Splicing—probably the oldest craft in the industry—was transformed. In the past, individual wires were stripped, wrapped together, and a sleeve placed over the repair. Now two wires were spliced by placing a newly developed sleeve over the wires and crimping it to connect the two wires. Contacts previously made by screwing a stripped wire onto a post were done faster by an inexpensive hand tool which forced the wire through a small groove in a pin, thereby making a contact quickly and without stripping the wire. The careful wrapping and soldering of wires on a frame was replaced by the "gun wrap."

Microchip technology increased the negative aspects of the already highly-automated industry. The general trend toward de-skilling continued, especially in the craft. Electronic switching allowed the company to transfer a number of craft workers' tasks to lower-paid clerical workers. For example, previously a subscriber was connected to the central office by a craft worker who would physically connect a wire; this task was now accomplished by a clerical worker via an electronic connection. Automated systems removed the worker's control over the quantity and quality of work performed. Work became fragmented into smaller, more meaningless, repetitive tasks. Operators no longer had the small satisfaction of completing a difficult connection or handling a variety of calls.

While automation of the company had not reached the stage of

The famous truck parade during the 1977-78 lockout.

the fully-automated electronic office, by the late seventies most of the workers at BC Telephone could see the writing on the wall. The next phase of the automation program would be the integration of the office record systems—CRIS, MAIS, and SOUL. Eventually a subscriber would be able to telephone a service representative and place an order for service. Through CRIS, the service representative would check the customer's service record and billing information. The representative would then use MAIS to locate the available facilities information, enter the new order into SOUL, and reprogram the central office computer with the new information. The entire process would be carried out by a single employee and be completed while the customer waited on the line. There would be a drastic reduction in jobs, and the skills and pay attached to the work would be reduced as well.

The central concern of workers faced with such changes is job security. Automation means that fewer workers can maintain and run the telephone network. The union had learned its lessons with dial conversion. For a period—while new systems are being installed, tested, and integrated—the work force will grow. But once the system is fully integrated, jobs disappear. The telephone workers had learned through bitter experience that the time to act on technological change was before the introduction of the change.

The union also sought to ensure that the existing work force would be trained in the use of the new technology. Skills which had taken years to acquire were becoming obsolete, and the new technology allowed BC Telephone to place skilled workers, such as computer technicians trained outside the company, directly into new craft jobs. In fact it became easier and cheaper for the company to do just this than to retrain its existing work force. The union demanded that the maintenance of telephone computing systems be done by existing employees. Work traditionally done by the telephone workers should continue to be done by them. Where new skills were required, the company should pay for the retraining. The union noted in the *Transmitter* that the problems of retraining were more threatening than ever before:

> The communications industry is entering a Technological Revolution. BC Tel, which has trained so many specialists, may soon have many employees with obsolete talents. Telephone Company tradesmen and women are not like those of other industries. A carpenter can easily practise his trade with another employer, but a telephone repeaterman, switchman, assigner or any one of a hundred other classifications will never enjoy this opportunity.
>
> It is not the Union's purpose to stop technological change, but rather to ensure that our trades people are not put in the unemployment ranks because their particular job skills have been by-passed by technology. We recognize that it is more economical, in the short run, for the Company to hire new employees at lesser rates of pay and train them for vacancies that evolve, however, we firmly believe the Company has an obligation to ensure that every one of the long-term employees who had a hand in making the Company a success are given every opportunity to be employed throughout their productive years.
>
> If BC Tel truly believes that people are its most important investment, proper job training and job relocation are small liabilities to retain these long term assets.[5]

The union's fight for a voice in the pension plan and improved benefits was directly related to its concern over technological change. Bill Clark, plant general secretary in the early seventies, explained the fight for better pensions by arguing that the union was seeking "income protection." "We are faced now," explained Clark in January, 1973, "with an accelerated program, introducing dedicated plant, electronic offices (local switching), computer type toll operations, communications satellites, various types of carrier systems imposed on local and trunk cables, and faster and cheaper methods of placing and splicing cable." In the face of rapid changes in the industry, the union hoped to improve the pension plan to allow early retirement for members. "Providing early retirement with decent benefits for older members," argued Clark, would help "protect young members against layoffs." The union wanted to share in the benefits of the new technology, and first in line were the retiring members who had contributed most of their productive years to the communications industry.[6]

Throughout the decade the company attempted to introduce arbitrary changes in work organization, work methods, job classification and equipment—changes it viewed as the right of management. An early program to restructure work was called the Performance Improvement Plan. The plan was one of a series of efficiency programs introduced by the company. The company planned to transfer a number of craft functions to the clerical division, thereby downgrading many of the higher-paid job categories. Some changes violated the collective agreement, but the company had little to lose in such confrontations. If union members contested the violations, they could grieve. Eventually the cases might end up in arbitration, but the worst that could happen was that the company would be forced to live by the collective agreement. On the other hand, the company was bound to win some of the arbitrations. It could thus use the arbitration process to roll back gains that the union had won at the bargaining table.[7]

This approach forced the union to respond in kind. Union members became frustrated with the endless grievances and arbitrations. Individual locals, groups of workers, and on occasion the union as a whole opted for more immediate resolution of disputes through walkouts, wildcats, and direct job action. By the late seventies industrial relations and bargaining between the union and the company had almost completely broken down.

In 1974 there were three major walkouts, all arising out of unilateral company actions. The first walkout was sparked by disciplinary action against employees' protests over the introduction of Phone Mart. Phone Mart was part of a major shift in work methods designed to save installation costs. Where in the past the company had sent out repair personnel to subscribers' homes, the public would now be obliged to pick up their own telephones for installation and bring faulty equipment into the company's commercial stores. It amounted to less service for the same money. In August, 1973 the company informed the union that it intended to establish a self-serve retail store called Phone Mart in Vancouver's densely-populated West End. This pilot project, Canada's first, was slated to open in March, 1974. In preparation for the Phone Mart, 22,000 residences in the West End were pre-wired with special plug-in telephone jacks.[8]

Phone Mart marked a profound change in the industry. Self-service would mean fewer workers. It also meant a decline in service. The company promised the union that "no layoffs will be made because of a reduction in the work force caused by Phone Mart." Wary of the company's assurances, the FTW began a campaign against the "non-service" provided by Phone Mart. Thousands of leaflets were distributed door to door and at shopping areas in the West End. "Phone Mart is a wrong number," claimed the leaflet. "It's a brand new non-service." Bumper stickers declaring, "Phone Mart in 74—Layoffs in 75" and "BC Tel's Phonemart—Profit to Yanks, BC Unemployment—Our Thanks!!!" appeared on workers' cars, shirts, and even company vehicles.[9]

In February, 67 plant workers were given one-and two-day suspensions for wearing the bumper stickers on their backs. The suspensions sparked a one-day walkout by a further 2,000 plant workers in sympathy with the suspended employees. The union defended the workers' right to express their opinion and grieved the suspensions. The union also took the Phone Mart project to arbitration claiming that it was a violation of the collective agreement. This arbitration, however, was lost when the chair and the company's nominee on the board ruled that the contract clause which stated that "any work having to do with maintenance, repair, alteration or construction of the telephone plant" must be done by qualified employees referred only to skilled work. Boris Mather, the union's nominee, disagreed, pointing out that the clause did not say "skilled work" but "any work." The union protested that BC Telephone should not be permitted to implement such significant changes as the Phone Mart without prior public hearings by the federal regulatory body, the Canadian Transport Commission (CTC). But the CTC did not hold hearings on BC Telephone until the fall of 1974, and those hearings were convened because the company was seeking a rate increase. The union's arbitration and publicity campaign had failed to stop the Phone Marts. The company declared the West End pilot project a success and opened a second store in Richmond a year later.[10]

Less than two months after the Phone Mart walkouts, 250 craft workers from Nanaimo Local 3 walked off the job in protest over the company's use of temporary employees—mostly college students—and the union's inaction on the issue. In the 1973 plant appendix to the contract, a clause limited company use of temporary employees to 5 per cent throughout the year and to an unspecified larger percentage during the summer in order to replace employees on vacation. The union assumed that temporary workers would not do craft work since the agreement specified that all maintenance, repair, alteration, or construction of telephone plant was to be done by qualified telephone journeymen, or by apprentices under the supervision of journeymen. The company, however, regarded the temporary employee clause as giving it the right to deploy temporary workers as management saw fit.[11]

Ron Jeeves, a leader of Local 3, told a newspaper reporter that if the company's interpretation of the agreement prevailed, all journeymen's positions could be made part-time, particularly with the growing use of highly-automated equipment. Jeeves maintained that the company was "making the journeymen climb all the poles and run all the wire while the student puts in the phone—the easy work." According to Jeeves, the company was misreading the intent of the vacation clause in the agreement by allowing part-time workers with twelve hours of classroom instruction to assume the easier jobs in the craft. "We cannot tolerate our permanent jobs, no matter what per cent of temporary employees there are, being taken over by non-union, non-trained . . . personnel," asserted Jeeves.[12]

The walkout lasted nearly a week and ended only after a special meeting of the plant council decided to fight the company's use of temporary workers in craft positions. The union decided to grieve and arbitrate any company attempts to use temporary workers as apprentices without meeting the training requirement. It also began to keep close watch on the use of trained apprentices; they were to be used only in accordance with the collective agreement.[13]

A few months after the Nanaimo craft workers' wildcat, clerical workers in Vancouver began sympathetic walkouts over a service centre clerk who was suspended when she refused to train a new employee without extra pay. The company claimed that the clerk was not training an employee but merely "familiarizing" the new worker. Under the clerical appendix to the collective agreement, the training of new employees fell under a separate job classification, that of service centre clerk III (training), which received a higher rate of pay. The union claimed that the company was trying to eliminate the higher classification by not promoting anyone to the position and by attempting to persuade service centre clerks to "familiarize" new employees with the job. At the time the company required more than 400 service centre clerks, all of whom had to be trained on the job. The company, however, had only three regular employees and one part-time employee in the senior training position. According to Chris Andersen, a business agent for the FTW, BC Telephone's refusal to pay training personnel at the higher rate, and the consequent suspension, amounted to "the company trying to gain things during the life of the contract that weren't originally negotiated."[14]

The walkouts spread, with eventually 400 clerical workers walking off the job. In face of the walkouts the company suggested that the issue go to binding arbitration. But the union refused, feeling that the company was setting a pattern of ignoring the collective agreement and stalling the union through long and expensive arbitrations. "The company has acted in deliberate and flagrant violation of the collective agreement," stated FTW president Bob Donnelly, "and we're not prepared to go to binding arbitration." On the evening of June 26, with the company refusing to back down on its suspension of the clerk, Donnelly called a province-wide walkout of telephone workers. The following day 10,000 telephone workers stayed off the job. Although the walkout occurred on a pay day, it was completely successful. Donnelly advised the workers to stay off the job until "such time as an amicable settlement can be reached."

No pickets were set up by the union during the wildcat walkout. The union claimed that pickets were a tactic for contract negotiations, not contract violations. The absence of pickets threw the company into confusion; companies usually ended walkouts by obtaining an injunction against picketing. By the end of the day the company had filed for an injunction against the union, alleging an illegal strike and conspiracy to damage the company.[16]

At the opening of the injunction hearing, Judge H. E. Hutcheon volunteered to act as binding arbitrator on the issue. When the company agreed to lift the clerk's suspension, the union accepted arbitration, ending the province-wide walkout. Judge Hutcheon upheld the union's contention that employees who trained new staff should be classified in a higher pay group. Donnelly felt that the union had been vindicated. "We feel rather regretful we had to take the action we did," explained Donnelly. "But unfortunately it was necessary to make the company realize they can't get away with their kind of games."[17]

In the fall of 1974, the CTC held hearings on BC Telephone's request for a rate increase. The FTW intervened in the hearings in an attempt to raise the issue of Phone Marts and job security. The union also opposed the company's proposed charge of 25 cents for directory assistance, which through decline in the use of the service would result in a reduction of approximately one hundred jobs. The union further objected to the company's application for a two-tier rate system with higher rates for those who did not use the Phone Marts.[18]

The main intervener in opposition to the proposed rate increases, however, was the provincial government. The NDP, which had been elected in August, 1972, was committed to public ownership of the telephone system. But shortly after the election, the new premier, Dave Barrett, stated that public ownership was a low priority. Instead, his minister of transport and communications, Robert Strachan, lobbied the federal government for the transfer of regulation over BC Telephone to the province. Strachan argued that provincial regulation would make the question of ownership "academic." But the provincial government hedged its position with a purchase—through the Civil Service Pension Fund—of more than $5 million worth of BC Telephone shares. This purchase totalled 3 per cent of the common shares and made the province the largest shareholder after GTE. The government hoped its shares would give it a seat on the board of directors and a window into the industry. In the spring of 1975, however, the provincial government's attempt to put its two nominees on the board of directors was soundly defeated by the massive GTE proxy.[19]

Barrett recognized that no provincial government could ignore as essential a utility as the telephone. So the government used the rate hearings to mount an attack on the telephone company's management and the problems inherent in vertically integrated multinational corporations. Intervening on behalf of provincial consumers, the government explained that BC Telephone was part of a multinational conglomerate. Because GTE was a vertically integrated corporation, its telephone companies bought most of their equipment from other GTE subsidiaries. The lion's share of buying took place within the conglomerate, with a significant mark-up at each stage. The government's lawyers argued that BC Telephone's priorities in purchasing equipment were determined by what other GTE subsidiaries produced, not necessarily the communication needs of the public. As a case in point, they noted that BC Telephone bought 82 per cent of its telephone equipment from two GTE subsidiaries, Automatic Electric (Canada) Ltd. and Lenkurt Electric (Canada) Ltd. They argued that BC Telephone decided in 1959 to go directly from step-by-step electromechanical equipment to electronic switching because Automatic Electric (AE) did not produce the intermediate generation of electromechanical equipment known as the cross-bar. Most of the other telephone companies in Canada converted to cross-bar switching equipment.[20]

The government also attacked the company on its quality of

service. Expert witnesses testified that "the service had deteriorated markedly between 1971 and 1974." A telecommunications consulting firm reported their findings from a one-day test study of quality of service during regular working hours. They found that 18 per cent of the long distance calls made in British Columbia were incomplete after 30 seconds; 22 per cent of the calls to the repair service were not answered within this time; and 16 per cent of calls to directory assistance were incomplete after the same period.[21]

The government's lawyers argued that the CTC should refuse the rate increase on the grounds that the company had not operated prudently, efficiently, and in the best interests of the public. Refusal of the application would serve as a stimulus for BC Telephone to increase its efficiency and effect cost savings in the public interest.[22]

The company defended its decision not to upgrade its switching equipment in the sixties by converting to cross-bar switching; it claimed that it had anticipated the eventual evolution towards electronic switching. The company admitted that it had expected electronic switching to have evolved faster than was the case, but the company denied that the decision not to go to cross-bar was related to the fact that GTE did not produce the equipment. The company's lawyer finally argued that the proposed increase was "a modest application that demonstrates maximum restraint at a time of inflationary pressures which have been difficult to predict."[23]

The CTC generally ignored the issue of quality of service. To the limited extent that it ever dealt with the problem, the CTC relied on telephone company testimony. The regulatory body did not draw up its own measures or criteria for quality of service. This hands-off policy extended to the issue of job loss, which the CTC declared to lie outside of its jurisdiction. As the CTC felt that the company should be free to exercise its managerial prerogatives as it saw fit, the commission focussed entirely on capital expenditures and projected revenues. If the company could satisfy the commission that its projected expenditures were reasonable and that its revenues were too low to pay for these expenditures, then the CTC would grant a request for a rate increase.

The criteria that the CTC used to judge a rate application mitigated against the arguments of both the union and the provincial government. It also created long-term problems for the union and for telephone subscribers in British Columbia. The CTC regulated in such a manner that it rewarded the replacement of people with machines. Expenses accrued by BC Telephone for purchase of new equipment and expansion of plant were easy to justify and contributed to the company's rising cost of operating expenses. As telephone equipment purchases were made almost entirely from Automatic Electric, these purchases increased the revenues of BC Telephone's parent, GTE. The more equipment BC Telephone bought, the greater its expenses and the stronger its case for rate increases. Telephone subscribers footed the rising bill for new expenditures while GTE profits grew through the purchase of equipment from its subsidiaries. The manufacturing side of the company was free from regulation.

Early in 1975 the company was granted its 4.6 per cent rate increase, to take effect in the middle of February. The provincial government announced that it would appeal the decision in the Supreme Court of Canada. The appeal, however, was withdrawn when federal minister of communications Gerard Pelletier agreed to an inquiry into the relationship between BC Telephone and GTE.[24]

In March, only 27 days after the company had been granted its rate increase, BC Telephone filed a new application for a 20 per cent across-the-board increase. Robert Strachan, the provincial minister of communications, called the application "shocking." Strachan pointed out that the company had reported a net profit of $50 million over the last two years and had just received a rate increase which added a further $15.3 million to its annual revenues. The company was particularly brazen, as the federal government had begun the inquiry into company payments to GTE, which totalled $83.7 million in 1973. "This action by BC Tel," protested Strachan, "shows only contempt for the public of British Columbia and the government of Canada. . . It's the kind of thing I suppose we must expect from a corporation that takes its instructions from New York."[25]

BC Telephone's president, Basil Beneteau, explained that the increase would net the company an additional $50 million in revenues to help meet rising operating costs and debt servicing charges. According to Beneteau, it would help the company attract outside capital in order to complete a $225-million capital works program in 1975. Beneteau warned that unless the company received the increase, the company would lay off between 2,000 and 2,500 employees. On April 27, BC Telephone began cutting staff by laying off 170 regular employees.[26]

The layoffs, the first since the sixties, took place while the union and the company were negotiating a new contract. Talks had begun in the fall of 1974, with the contract due to expire on December 31, 1974. Following the established pattern, negotiations broke off in December, with the union calling for the intervention of a conciliation officer. The conciliation officer booked out of the dispute in January, and Vancouver lawyer Hugh G. Ladner was again appointed conciliation commissioner. Near the end of February Ladner submitted his non-binding report.[27]

The union bargaining team was divided on its recommendation to the three divisions. Plant and traffic bargainers favored acceptance, but the clerical leadership recommended a rejection vote. Clerical felt that Ladner had not dealt adequately with job classifications. As well, the clerical division had sought an across-the-board settlement and a substantial raise in the base rate for clerical workers. But the leaders failed to win over the clerical workers, who went along with plant and traffic and accepted the Ladner report.[28]

The proposed contract included a wage increase, of 18.5 per cent effective January 1, 1975, and a further 14 per cent increase in 1976. It also included a jointly-trusteed bargaining unit employees' pension plan similar to that established at OK Telephone. New clauses dealing specifically with technological change were added to the layoff provisions of the contract. These included twelve months' notice of layoffs due to technological change, and the company was to pay all costs of retraining and relocation of any employee affected by technological change. Employees displaced by technological change who accepted a job in a lower classification were guaranteed that their wage rate would be maintained. The union also established a reduction in the worktime through a clause on accumulated time off, whereby an employee would accumulate additional paid leave of half a day each month. This reduced the work year by six days. A further public holiday, British Columbia Day, was added to the list of holidays.[29]

The company accepted the Ladner report in early April 1975, but refused to sign the agreement until it learned the outcome of a clerical classification arbitration which affected 50 business accounts representatives. Union leaders suspected that the company was trying to provoke a walkout. The union organized demonstrations outside company offices to protest the delay but refrained from any further actions. Finally, more than two months after agreement had been reached, the contract was signed on June 18, 1975.[30]

In addition to negotiating the new contract, the union and the company met to discuss the layoffs. The union linked the layoffs to the company's new rate application before the CTC. The company, however, told the union that the layoffs were necessary for economic reasons. A few months earlier the company had threatened to lay off operators because of "a 25 per cent reduction in operator handled long distance calls." But the April layoffs were confined largely to the plant division. The union protested that workers were being used to extort a rate increase from the CTC.

They noted the parallels between the company's actions and a similar tactic taken by Bell Canada. Bell, like BC Telephone, had an application before the CTC and had threatened massive layoffs and the cancellation of its construction program if its request was not granted.[31]

The CTC granted BC Telephone an interim 10 per cent rate increase, and the laid-off workers were recalled. Further talk of layoffs ceased, but the threat had lasting repercussions. As a multimillion dollar business with more than 13,000 employees, the company's laying off of fewer than 200 workers for a few months had a minimal effect on its expenses. However, the effect on its work force, particularly the new generation of workers, recruited since the last layoffs in the sixties, was serious. They began to question the sincerity of the company promise of a good job and a life-long career.

As late as August, 1974, the company had run an advertisement in the Vancouver *Sun* to attract new workers:

> **WANTED**
> **HELP**
>
> Your telephone company really needs clerical help. So, if you really want a job, we might just have one for you. My name is Anne, and I work as an employment representative. Honestly, I didn't expect to appear in this ad, but I can tell you from personal experience that BC Tel is a good company. Good pay. Good benefits. A chance at a career. Or a chance to start again in a place where maybe, but maybe (with all our different offices) you'll be working closer to home. Naturally we'd like experience. Typing. Filing. Business machines. Or general office work. But, we're big enough to train you. And there's a variety of jobs going. If you're interested, we're interested. Drop in to see us at the employment office—856 Homer Street. Usually our ads say, "You can talk with us." And you can. But this one's special. So we're saying . . . "You can work with us."[32]

Less than a year later the company was threatening to lay off the very people it had enticed through its publicity campaign.

The layoffs in 1975 also revealed a major problem with the new technological change clause in the contract. The company could ignore the clause by invoking economic justification. The onus was on the union to prove that layoffs were caused by technological change and not economics. Exactly how it could do this was far from clear. While it was obvious to the union that the reduction in operator-handled long distance calls was the result of direct distance dialing made possible by the TSPS, the company could claim economic factors as the grounds for the layoff.

According to George Yawrenko, plant general secretary, the layoffs "resulted in all kinds of arbitrations because the company tried to retain people who they felt had special qualifications." The company wanted to lay off people in certain job classifications in some areas of the province but not in others. The FTW, however, forced the company to adhere strictly to seniority.[33]

An important arbitration case dealt with the demotion of supervisors. Of the 170 employees laid off, 160 were in plant and included 152 classified employees and eight supervisors. The company instead of laying off the supervisors attempted to demote them to the craft ranks. The plant division refused to accept the supervisors and took the case to arbitration. The arbitrator sided with the union, ruling that "the company is not entitled to return eight supervisors to the bargaining unit during a period of layoffs."[34]

In spite of a clause in the collective agreement which stated that arbitration decisions were to be "final and binding," the company appealed the ruling in the Supreme Court of British Columbia. The court, however, upheld the union's contention that supervisors could not be demoted into the bargaining unit. The demotion of a supervisor constituted an addition to staff, which was not permitted during layoffs, according to the collective agreement. The company's argument that the supervisors were senior employees carried no weight; employees' seniority rights were ruled to arise from seniority within the union, not total time with the company.[34]

The exclusion of supervisors and other personnel from the bargaining unit had, since the late sixties, become a major point of contention between the union and the company. In this period the classifications outside of the bargaining unit had increased sharply. The company constantly attempted to reassign work to the excluded classifications, and the practice had become a major threat to union members' job security. The two strikes had demonstrated just how effective the growing number of excluded workers could be in weakening the union's principal weapon—the strike. With ongoing technological change, the union feared that the bargaining unit would continue to shrink relative to the excluded personnel, until the union was destroyed. Over a period of 30 years the ratio of bargaining unit to non-bargaining unit personnel at BC Telephone had shrunk from approximately 19 to 1 in 1949 to 3.7 to 1 in 1979.[35]

A huge increase in exclusions had occurred since the mid-sixties. The union's attempts to stop this erosion of the bargaining unit were impeded by contradictory directions and decisions from the Canada Labour Relations Board (CLRB). Beginning in the late sixties and continuing through the seventies, the FTW was involved in an extended battle with the CLRB over the definition of the union's bargaining unit. The board condoned almost every company exclusion from the bargaining unit and tied the union's hands over the prevention of further exclusions.

The roots of the battle can be traced to the late forties, when the FTW discovered that its legal recognition as bargaining agent for employees at BC Telephone, granted under PC 1003, was in jeopardy. As an emergency wartime legislation, PC 1003 was abolished in 1948. Most provinces then enacted labour legislation to allow union certification and the continuation of the industrial relations machinery established under PC 1003. But workers at BC Telephone were not eligible for certification by the province. As BC Telephone fell under federal regulation, industrial relations at the company came under the jurisdiction of the federal Department of Labour.

In February, 1949, the CLRB certified the FTW as bargaining agent at BC Telephone, excluding 300 employees, 8 per cent of the work force of 4,000. Excluded from the bargaining unit were 10 executives, 22 confidential employees, 30 professional employees and 261 supervisors. Each job excluded from the bargaining unit was listed. There were problems inherent in this listing. The investigating officer who drew up the list of "occupational exclusions" suggested that the description of the bargaining unit was not "sufficiently precise." Change was constant in the industry and new job classifications or significant changes in old jobs would make the existing delineations obsolete. The CLRB explained how it envisioned such a problem would be resolved:

> You will realize that the Board, when issuing a certificate, can take account only of the classifications existing at the time of the investigation. Subsequent changes in title which involve no changes in duties should be settled easily by the parties themselves within the framework of the Board's certification. In cases involving entirely new classifications or drastically revised duties, where the parties could reach no agreement, the matter would require consideration by the Board.[36]

Essentially the union's certification was inclusive, with all jobs except the specifically-named 187 job classifications included in the bargaining unit. But by the mid-sixties two significant transformation had taken place. The company had completed a massive technological changeover to dial, and the company was now owned by GTE.

Early in the sixties the union signed up most of the technicians and succeeded in adding the technician's classification to the plant

Leigh Taylor

Picketing the Seymour Street BC Tel building during the 1977-78 lockout.

appendix through negotiations with the company. Similarly, the union added sales personnel to the bargaining unit in the same period. But midway through the decade the company began to create new classifications which the company unilaterally declared to be outside of the bargaining unit. These included analysts, training personnel, data consultants, and marketing and planning personnel. As a general rule the company designated any collecting and organizing of data as a managerial function. With the growth of computer technology, these exclusions from the bargaining unit threatened the clerical division.[37]

The union attempted to prevent the company from creating these new positions and assigning traditional bargaining unit work to non-bargaining unit personnel. The company's creation of non-bargaining unit jobs was a fundamental threat to the union. When the union approached the CLRB over the company's unilateral exclusions, the board declared the issue to be one of support, not of work jurisdiction. The decision meant that the union faced the uphill battle of trying to persuade employees hired in new job classifications—who according to the company were supervisory or management employees—to join the union. In the past, new or promoted employees would automatically remain in the bargaining unit as long as their work was customary work of the bargaining unit. With the new ruling, the company could now create new positions which included bargaining unit work but call these positions "management jobs" and place them outside of the union. The onus was now on the union to sign up a majority of these members and appeal to the board for a change in the union's certification. What had been an inclusive bargaining unit was now exclusive. As an inclusive unit the burden was on the company to prove that new classifications dealt with work outside of the union's jurisdiction. The CLRB had reversed the union's certification. Bill Clark, the plant general secretary, called the board's action "a complete sell-out to the employer and the professional groups."[38]

A second option open to the union was arbitration. The union could submit to arbitration the new classification as an attempt to remove work from the bargaining unit. Arbitration, of course, is a costly and slow process, and does not always attain the desired results. The union, for instance, submitted to arbitration the creation of a new position called "communications consultant–data" in the early sixties. However, after the union won the arbitration, the company dropped the classification. This action convinced the union that arbitration of each new position would be a costly route which the company could frustrate whenever the union was successful.[39]

During the 1973 contract negotiations the union made headway on the issue of exclusion of classifications from the bargaining unit. The conciliation chairperson in the 1973 dispute, Hugh Ladner, recommended a procedure whereby the union and company would mutually work out which jobs would be bargaining unit and which would be excluded. Where the two parties differed, the issue would be submitted to an arbitrator. After agreement, the parties would make a joint application to the CLRB for an amendment to the union's certification. But shortly after the procedure had been agreed upon, it had to be scrapped because the CLRB informed the union and company that only the board had the right to define a bargaining unit.[40]

The issue was further complicated in March, 1973, when the federal labour code was amended to allow supervisory and professional employees collective bargaining rights. The two strikes and the numerous walkouts during which the supervisors had maintained service had created a barrier between the union and this group of employees, frustrating attempts by the union to organize the supervisors. The company, well aware of the non-unionized employees' vital role in maintaining service during a strike, encouraged the division.

Like the central office craft workers of the twenties, supervisors saw themselves as privileged. They expected special benefits and resented the union, which they saw as seeking to destroy their special status. But a number of the supervisors also recognized that as non-union employees they lacked basic job protection. The change in the federal legislation gave impetus to a growing movement among supervisors to organize.

In the sixties professional engineers and plant supervisors formed associations which the company refused to recognize. With the change in legislation in May, 1974, a group of engineering department supervisors formed the Society of Telephone Engineers and Managers (STEM) and applied for certification. A month later a group of first-line—the rank closest to the shop floor—plant supervisors formed the Telephone Supervisors' Association (TSA) and applied for certification.

The plant department foremen were particularly interested in organizing. Most employees at the first-line level in plant were former bargaining unit employees who had been promoted to foremen. In most craft unions these jobs fell within the bargaining unit. The TSA's attempt to win certification failed as the CLRB ruled that plant first-line supervisors alone were not an appropriate bargaining unit.[41]

Early in 1975 the FTW applied to the CLRB for a broadening of its certificate to include those employees now eligible to organize on the basis of the new legislation. But this application was dismissed, with the CLRB ruling that the union needed to show majority support of the supervisory employees. In 1978 STEM and TSA merged to form the Telecommunications Employees Managerial and Professional Organization (TEMPO). The CLRB declared 2,075 management personnel to be "employees" and eligible to organize. The board gave TEMPO a certification vote, which failed to win the required majority, with 896 employees voting in favor and 975 against.[42]

By the late seventies the basis of future confrontations had been established. The union's pursuit of job security challenged the rights of management as the company defined them. Also, one might add, the CLRB and the CTC seemed to accept the company's view. Throughout the decade job security had been under attack. Technological change, reclassification and reassignment of work, and contracting out all undermined job security. The union's attempts to enforce the original inclusive certification had resulted in the CLRB reversal of the original intent of the document. The union's appeal to the federal regulatory body to stop the Phone Marts, which caused a deterioration in service to customers and threatened jobs, was rejected because the CTC ignored the issues of quality of service and job loss. The only effective tool that the union had in its fight for job security was the collective actions of its members. And job security meant the very life of the union.

Chapter 9 *The Long Distance Feeling*

In the spring of 1977 the Federation of Telephone Workers of British Columbia became the Telecommunications Workers Union (TWU). The telephone industry had evolved from basic telephone service into the wider field of telecommunications. The union's change of name paralleled constitutional and structural changes aimed at consolidating and uniting the union in order to better respond to the new challenges posed by management and the industry.[1]

The FTW had always been an anomaly. Legally it was one union; it had one certification at BC Telephone. Yet it functioned as three separate unions. Members paid dues to the division of the union to which they belonged. Not only had each division a different dues structure, but the division appendices to the contract were as diverse in content as three separate agreements. Each division paid an assessment to the central administrative fund, which paid the wages of the office staff and office expenses. In practice the union was three separate organizations in a loosely knit federation. They shared common office space and collaborated on common issues which affected all three. There were three chiefs—the three general secretaries—each answerable only to their own division, and each had one or more assistant general secretaries working with them.

Telephone workers joined the FTW by applying to the appropriate local based on their place of work and the type of work they did. Generally each area had three FTW locals—traffic, plant, and clerical. The authority to charter new locals rested with the divisions. At one point the divisions even required different majorities in strike votes.

The FTW held no general conventions. Instead each division held yearly council meetings. Each local elected at least one councillor, who served as that local's delegate to the division council. Division councils met separately, with interdivision liaison taking place only through the federation council. The federation council was an executive body which met only twice yearly and comprised the union president, the three general secretaries, and two members from each of the division councils. Although there was collaboration between the general secretaries, this was largely informal. Each division was constitutionally autonomous.

By the early seventies it was clear that the structure of the union was a source of weakness. Internal communications in the union were abysmal. With no formal mechanism for the divisions to meet, there were frequent misunderstandings between the divisions and no avenue for resolving problems, as well as long delays in implementing policy. The union's divisional structure had become a major barrier between members of different divisions.

The 1969 strike and subsequent job actions brought members together for the first time. The implementation of technological change also reduced the distinctions between plant, traffic and clerical work. Changes introduced in one work area generally had repercussions in others. The rapidly changing technology in the telecommunications field and the need to respond immediately to company-initiated changes made it absolutely necessary that the three divisions work in close cooperation.

The plant division initiated many of the early proposals for change. The membership distribution within the union had changed considerably since the union was first organized; the plant division had increased both relatively and absolutely. In the forties the traffic division was the largest division, with plant second and clerical a tiny third. However, by the seventies plant was the largest division, slightly larger than the other two divisions combined. Clerical membership now outstripped that of traffic, which had been decimated by automation.

In the early sixties growing militancy and increased dissatisfaction with the response of the FTW leadership to changes in the telephone industry led to a strong movement for structural change. This demand coincided with the growth of the plant division and the decline of the traffic division. However, the CWA raid had put the union on the defensive, and the FTW leadership viewed as suspect any discussion of structural change, which implied criticism of the existing structure.

The 1969 strike reinforced the desire for change. After the strike the federation council received a number of resolutions calling for a restructuring of the union and the abolition of the divisions. But fear over the loss of divisional autonomy led to the defeat of these resolutions. The divisions had different methods of deciding their negotiating priorities. In the plant and traffic divisions the councils established bargaining goals; in the clerical division the council set the general priorities, but the bargaining team decided the specific items for negotiation. Fear of upsetting tried practice in bargaining, and the possibility that divisional concerns would be ignored, led to the postponement of changes. But the union was carried along by circumstances. The lengthy OK Telephone strike gave further impetus to the move towards a single union. FTW president Gordon Cooper worked full-time for the union through most of 1973, leading the union in two drawn-out sets of negotiations, first with BC Telephone and then with OK Telephone. The company in fact asked the union to place Cooper on staff because his union duties made it impossible to plan a work schedule for him. During the long Okanagan strike the union realized that its principal officer should be available full-time. In January, 1974, the federation council voted to make the union presidency a full-time position.

In the spring Bob Donnelly replaced Cooper. Originally from Scotland, Donnelly immigrated to Vancouver as a teenager. He started work at BC Telephone in 1957 as an apprentice in the central office. In the mid-sixties Donnelly transferred to PBX repair and maintenance and in 1968 became an engineering technician. Donnelly had been a long-time activist in Vancouver Local 1, holding a variety of executive positions in the local and representing the local on the plant council. He had worked as assistant general secretary to both Bert Johns and Bill Clark. When Clark suffered a heart attack at the start of the OK Telephone strike, Donnelly took over the responsibilities of general secretary. Clark, on his return to work in the fall of 1973, resigned from his union office and returned to the tools. In January, 1974, Donnelly was elected president.[2]

George Yawrenko replaced Donnelly as plant general secretary. Yawrenko began with BC Telephone as a lineman in Abbotsford in 1951. He was a leader of Fraser Valley Plant Local 32 and had served as its councillor since 1965. In 1972, when the plant division decided to elect an assistant general secretary to serve the interior, Yawrenko was elected to the position. The plant division opened an Okanagan office in Vernon that summer. Yawrenko was the only full-time union officer in the Okanagan during the 1973 strike. On Yawrenko's election as plant general secretary, Gordon Stoutenberg, a craft worker from the Okanagan, became assistant general secretary for the interior.

Donnelly's election to the presidency marked an important change. He was the first president of the union to become chief executive after holding the powerful position of general secretary. Donnelly made the move with the intention of transforming the role of the chief executive. He saw this as a necessary step in the move toward one union. Previously the president had been largely a figurehead; the president's main responsibility was to chair federation council meetings. Though the position had become a full-time job, it carried little authority. The power remained with the divisions and their general secretaries.

Shortly after Donnelly's election as president, the federation council struck a committee to make recommendations on constitutional changes. These recommendations were adopted by the council in the summer of 1975. In February, 1976, after the membership ratified the proposals, the council began to implement the changes. The divisions were not abolished, but they lost their autonomy. The general secretaries became vice-presidents of the union and the assistant general secretaries became business agents. The role of the secretary treasurer too was substantially changed; while in the past it had been a stenographer and bookkeeper's job, the secretary treasurer now became a staff officer with Tom Stanley, a craft worker, elected to the position. With a full-time president and secretary treasurer responsible for the records and finances of the union, the union was beginning to centralize. The division councils continued to hold annual meetings. But in addition the new constitution established an annual TWU convention to be attended by councillors from the three divisions. In compliance with these provisions the first convention of the Telecommunications Workers Union was held on June 12 and 13, 1977.

At the time of the founding convention the union was again in a bargaining stalemate with the company. The agreement had expired at the end of the previous year. Although negotiations had begun months before the expiry, few issues were being resolved; the talks were strained. The federal government had introduced wage and price controls and the company had decided to take advantage of the climate to demand concessions from the union.

The Anti-Inflation Act (C-73) was introduced in 1975 under the guise of combating inflation. It severely restricted the bargaining power of workers by limiting wage increases. Prices were supposed to be controlled and rise only in relation to costs, but land prices,

Leigh Taylor

TWU president Bob Donnelly at the Christmas party outside the Boot, December 1977.

interest rates, basic food prices and energy costs were exempted from the controls program. These were all major components of the consumer price index. The legislation made no attempt whatsoever to limit profits. The absence of firm controls on prices meant that while workers' wages were forcibly restrained, prices continued to rise. The labour movement labelled the Anti-Inflation Board (AIB) controls program "Wage and Wage controls."[3]

The AIB marked a major intervention by the government in the collective bargaining process on the side of employers. By tying the hands of labour on money items, the government left labour vulnerable to further attacks by employers. Employers used the restraint program to squeeze concessions on non-money items from the labour movement. With little money on the table, and the knowledge that no more could be granted, labour leaders were hesitant to go to their membership for a strike vote. Members were hesitant to go on strike over what seemed to be insignificant changes in a few clauses in the collective agreement.

BC Telephone tried to take advantage of this bargaining climate to eliminate the contracting-out provision in the TWU contract and to add a management rights clause. The contracting out and management rights issues were both linked to technological change and job security.

The contracting-out clause limited the company's right to purchase goods and services from outside firms. In the sixties and seventies the company removed a variety of jobs from the jurisdiction of telephone workers. Until the early sixties, cafeteria staff had been organized by the telephone workers. But with the contracting out of this work to an industrial caterer, the cafeteria staff was removed from the union. Likewise, until the sixties, the company had done most of its own construction work. Construction jobs were generally entry level positions for craft workers. Throughout the decade more and more construction work was contracted to outside firms.

The introduction of computers made it possible for BC Telephone to further extend its attack on union jurisdiction. While the maintenance of central office equipment remained craft work, as more of this equipment became computerized, the company claimed that computer repair was not traditional craft work. The restrictive contracting-out clause, however, meant that the union could ensure that its members continued to service and be retrained in the use of the new equipment. Without the contracting-out clause, computer repairs and maintenance could be contracted out to computer firms, outside technicians, or anyone else the company chose.

The contracting-out clause in the expired agreement listed more than 30 jobs which the company could contract out:

> The following jobs may be contracted out without notification to the Federation of Telephone Workers: snow clearing, road grading, road improvements, window cleaning, gardening, garbage disposal, cleaning manholes, training (courses not available through the training centre), furniture houses—supply and assembly furniture, cartage of major equipment (such as PABX's), engraving, traffic control, rockholes, backhoes and operator, major underground digging—placing duct, pouring cement, brushing, butt treatment and inspections, setting poles by helicopter, submarine cable placing and maintenance—other than splicing and fault finding, tramway operations, motor vehicle—repairs, maintenance, and servicing of leased vehicles, towing for repairs and vehicles loaned to another area not accompanied by the assigned employee, CPR open wire, chrome and nickel plating (coin telephones), coin booth cleaning, coin booth and PABX Electrical AC, diesel generators—major overhauls, construction of manholes, major building construction, cafeteria, guard services, meter reading beyond a fifteen mile radius of the normal work location of two or more Plant Clerks, cable ploughing (operator only).

In addition to these exclusions other jobs could be contracted out, but the union had reasonable protection in that the collective agreement required the company to negotiate with the union any further contracting out. If the two parties were unable to come to an agreement, the contract provided for a single arbitrator to decide the issue within 72 hours.

The company had not fared well in arbitration over contracting out. It was therefore now trying to eliminate the clause altogether. By the mid-seventies the union had stepped up its vigilance against endless company attempts to contract out. The union began to enforce the contracting out provision in areas where it had been lax in the past.

One such dispute broke out in February, 1976, when the company sold miles of copper wire and hundreds of cross-arms along an abandoned line in the Nelson-Trail area to a salvage company. The salvage contract had an "as-is, where-is" stipulation. Five telephone craft workers were suspended when they refused to take down the wire and cross-arms and leave the material on the ground for salvagers to roll up and cart away. The union members argued that it was their job to coil, bundle, and pick up the wire in the field. The union insisted that BC Telephone employees transport the sold equipment to a telephone company depot, where it could be collected by the purchasing salvage firm.

The suspension of the five workers sparked a walkout of more than 150 telephone workers in the Nelson-Trail area. After two days the five workers were reinstated and the walkout ended. The union then sent the issue to arbitration. The majority of the arbitration board sided with the union, arguing that the company had violated the agreement by having telephone plant removed by persons other than plant employees. This arbitration award, typical of a number of decisions during the period, found unequivocally in favor of the union. The specific and detailed language of the contracting-out clause was a powerful aid in proving the union's case.[4]

The company saw the clause as a block to its right to manage as it saw fit; the clause became central to the company's list of bargaining concessions. The company offered to replace the clause with a job security and continuous employment clause. The new chairman and chief executive officer of BC Telephone, Gordon MacFarlane, assured the union that under the new clause no employee with two years or more seniority would "lose his or her employment as a result of technological change or contracting out from now until normal retirement age is reached."[5]

Bob Donnelly, president of the TWU and chief negotiator, pointed out the obvious loophole in the proposal. "They're guaranteeing full employment until they lay us off for economic reasons," countered Donnelly. "After they've contracted out all our jobs. . . it will be uneconomic to keep everyone on, and people will be laid off."[6]

Other concessions that the company sought in the 1977 negotiations included several items which the company had tried to introduce during the life of the last two agreements and which the union had successfully stopped through walkouts and arbitration. BC Telephone demanded that seniority be changed so that supervisors could continue to accrue union seniority. The company assured the union that it would not "demote managers to the classified ranks." But the union could not accept the company's assurances. During the 1975 layoffs the company demoted managers and attempted to place them in the bargaining unit. If BC Telephone did not plan to move supervisors into classified positions, reasoned the union, then why try to establish supervisors' union seniority?[7]

The company also attacked the contract provision which prevented temporary employees in the plant division from working on their own. The union had won an arbitration on temporary workers in the plant division after the Nanaimo walkout. The arbitrator declared that the company must follow the craft provision of the collective agreement which stated that all craft work was to be done by journeymen, or apprentices under the supervision of a journeyman. The company still hoped to use student and temporary workers to replace craft workers at a fraction of the wage cost. Having lost the issue in arbitration, it now sought to win the point in negotiations.

Negotiations broke down early in 1977, and conciliation officer Doug Cameron entered the dispute toward the end of January. Cameron only met with the parties for two days. He then booked out of the dispute and submitted his report. Early in February Dr. Noel Hall was appointed conciliation commissioner and met with both parties for over two months. By the end of May Hall submitted his report to the minister of labour and copies were sent to the company and the union for study.

Hall described the breakdown in bargaining between the TWU and the telephone company. "There exists a fundamental lack of mutual trust between the parties," wrote Hall. "Even the most minor change in contract language proposed by one party seems to trigger a predictable negative attitude or response from the other." Hall noted with distress that when he entered the dispute, "almost every clause in the collective agreement was up for either major or minor modification." The parties were moving in opposite directions, with the company "seeking greater freedom for managerial decision making" and the union "seeking further restrictions" on the company's decision-making power.[8]

Hall felt that mistrust and concern over technological change were at the root of the breakdown in bargaining. He chastised management for its refusal to work with the union on the issue of technological change. "It is apparent that the industry is on the threshold of major and far-reaching technological change that will have a serious impact on the levels of employment in various segments of the operation and that will have a major impact on the levels and kinds of skills required of the work force and which will affect opportunities for individual growth and development," explained Hall. "It is apparent that these impending changes and their potential impact have not been effectively defined by management and communicated to the employee group and their Union leaders, so that meaningful accommodations can take place through the collective bargaining relationship." Hall recommended that the contracting-out clause be left intact and that the union and company establish a joint study committee on technological change. "What is needed now," emphasized Hall, "is a thorough and systematic study of the issue, in which the union is involved from the beginning as a full and equal partner, so that the union and its members can have full confidence in the results . . . only then can the parties sit down at the bargaining table to reach new accommodations in this vital area."

On the wage issue Hall recommended an across-the-board increase of 45 cents an hour, which roughly equalled the 6 per cent increase permitted under wage control legislation.

A number of disputes led to walkouts during the conciliation and bargaining. They were sparked, for the most part, by company attempts to contract out plant work. By early 1977 the union had eighteen incidents in arbitration, many of which went to arbitration only after walkouts or other protest actions by employees.

In April a dispute broke out over the company's purchase of iron support frames from Northern Telecom Ltd. BC Telephone had traditionally fabricated its own frames through its subsidiary CT and S. But in the mid-seventies the company began to purchase prefabricated distributing frames from Northern Telecom. The union protested this action as contracting out and the issue had gone to arbitration. The company argued that while the basic "iron work" involved in making the frames had remained "essentially unchanged since the first known patent was granted in the United States in 1893," the unique Northern Telecom configuration represented a marked improvement in frame design. Northern Telecom frames, the company contended, provided greater economy of space, greater flexibility, better capacity for increased loads, and as they came largely assembled, it meant a substantial saving in installation time and expense. The union countered that while the Northern Telecom frames were an improvement, they were not qualitatively different from the basic iron work done by CT and S.

The arbitrator, R. Neil Monroe, handed down an ambiguous ruling. To the question posed in the arbitration, "Is BC Telephone Company entitled to purchase distributing frame equipment from any source available?" Monroe answered, "Originally yes, now no . . . the original purchase was not really contracting out . . . because CT and S did not and does not design or have the research or technology, nor the intent to create, such an improved frame complex" as Northern Telecom had developed. But now that the company had purchased frames, according to Monroe, the company should make a "reasonable effort" to have CT and S fabricate similar frames.[9]

Both parties interpreted the award as backing their stance. Through April and May the union organized rotating walkouts to protest the continued use of Northern Telecom frames and the company's refusal to resume fabrication of the frames in-house. Walkouts over the issue contributed significantly to the workers' growing militancy over violations of the security clauses in the contract.

By early June all attention turned to the Hall report and its recommendations. Shortly after receiving the report, the union accepted it. But the company was particularly displeased with Hall's recommendation to retain the contracting-out clause, and rejected the report. The company proposed that the clause be deleted before the union and company set up a committee to study the issue of technological change. The company suggested to the union that the Hall report be used as "a means of getting back to the bargaining table." But Donnelly, speaking for the union, noted that the union and company had "met face to face for eight months without resolving anything . . . we're certainly not about to negotiate all of a sudden with the Hall recommendations merely as a basis for more negotiations."[10]

The union did meet with the company again in July, but negotiations broke off almost immediately, with neither party willing to move. In August the union took a strike vote. Members voted 73 per cent in favor of strike action at BC Telephone and 90 per cent in favor at CT and S. After the strike vote tally the union met one last time with the company for nineteen minutes, with no results.[11]

The union and company prepared for a confrontation. The union announced that it was not planning an all-out walkout. Its strategy was to take innovative action, including demonstrations and other attention-grabbing protests. The "various skirmishes and harassment tactics" would allow the union to pressure the company without going into a full-scale strike. The union officials wanted to maximize the economic pressure on the company while

Leigh Taylor

Mass picketing during the 1977-78 lockout.

minimizing employee wage losses and subscriber inconvenience.

On Tuesday, September 20, the union announced that it was starting a work-to-rule campaign. The following day, catching everyone by surprise, the union held a spectacular protest in the form of a truck parade. More than four hundred company trucks converged on downtown Vancouver during the morning rush hour and slowly circled the BC Telephone building at 768 Seymour Street. The demonstrators drove around the block in tight formation for about half an hour with their lights and flashers on and their horns blasting. The parading trucks sported such signs as "Dr. Hall says it all," "Keep Jobs in BC," "Contract In Not Out," and a few relics from past disputes including the slogan "Phone Marts in 1974—Lay offs in 1975."[12]

The parade made national news, bringing the union's protest to the attention of the entire country. The company was furious. Company officials stood on the sidewalk during the parade and occasionally ran up to trucks and tore off their signs. Company officials copied down truck numbers of participants and assured reporters that all identified protesters would be docked a day's wage. Although the drivers returned to work after the brief parade, the company nevertheless docked the participants a full day's wage.

At the end of September operators held a brief protest at 768 Seymour Street by "downing cords" for fifteen minutes. The following day the clerical computer staff at the company's main computer centre walked out for 24 hours. On October 3 the union picketed the company's Regent, Amherst, and Richmond offices in the Lower Mainland. The following day BC Telephone's new headquarters on Kingsway in Burnaby was hit by a rotating strike, with more than 1,400 employees walking off the job.

The company retaliated against the walkout by locking out the workers. Signs declaring "no work" were placed on the doors of company buildings and workers were told not to return to work until the company called them back. A few days later the rotating strike hit 768 Seymour Street. Again the company responded with a lockout. By the second week in October more than 2,600 workers were locked out in the Lower Mainland and Victoria.

On October 13 the new traffic vice-president, Theresa North, marshalled a group of about 75 operators outside the Seymour building early in the morning. They stormed into the building, shoved aside the security guards, and marched up several flights of stairs to the operating rooms. The operators walked from one operating room to the next, observing supervisors and harassing them. According to North, the personnel staffing the boards, which included supervisors as well as the company doctor and nurses, seemed embarrassed. The locked-out operators, on the other hand, were furious as they witnessed supervisors and non-bargaining unit personnel doing their jobs. North commented that the "walk-through," as it became known, meant a lot to the workers. "Traffic felt really good about it," explained North. "It was really something for the telephone operators to take as militant an action as this."[13]

The day after the walk-through the company ended its lockout and called on the union to "enact a complete return to work." This led to a brief return to the bargaining table. After negotiations once again broke down, the union resumed its rotating strikes and the company its retaliatory lockouts.[14]

By November 23, the union had nearly 3,000 members off the job. The union executive planned no further escalations and intended to maintain this number of employees off the job. But many workers found conditions intolerable. Donnelly explained that "the membership has been under extreme pressure from management employees on the job and we finally felt any continuation of our control to keep them at work was no longer feasible." At noon on November 24 the union called out the remaining 7,000 workers, who were promptly locked out. "The union regrets the necessity of this action," commented Donnelly, "but BC Telephone has been harassing, intimidating and provoking people to the point where being locked out is preferable to working under these conditions."[15]

A few days after the outbreak of the full-scale lockout the company published a full page advertisement in the *Province*. Bordering the advertisement were 9,165 names of employees with two years or more seniority with the company. The advertisement explained to employees that "You will not lose your employment because of contracting out or technological change now or in the future." Company chairman Gordon MacFarlane supplemented the guarantee by adding in an interview that "it would take virtual economic disaster before we would lay off any employees." A few days later a similar advertisement appeared carrying 971 names of employees of CT and S.[16]

Donnelly responded by stating that the company was "misrepresenting the facts" and the union had not asked for "this stupid guarantee. . .they're guaranteeing employment only as long as it is economic and with their policy of profit before people that won't be long . . . What happens to our jobs," he asked, "after they've been contracted out? . . . All the union wants is the retention of the current contract language, which restricts the number of jobs that can be contracted out."[16]

The company and the union had different memories of the past decades. Company spokesperson Keith Matthews told reporters that when the company had introduced dial service, the union had fought the change on the grounds that jobs would be lost. Contrary to the union's contention, Matthews stated, jobs were not lost. Yet the company's own annual reports indicated that there were 7,130 employees at BC Telephone in 1957 and five years later, in 1962, there were under 5,330. Union members simply did not believe the company when Matthews assured reporters that the company did not see "any lessening of the employees. . .in the next few years."[18]

The full-scale lockout was the climax of a long dispute. The TWU did not want the lockout. The union was on the defensive, faced with a company demand which would ultimately lead to the destruction of the union. While union members were united, they felt isolated. Although the confrontation had been forced on the union, it had great difficulty in getting its position across to the public. The company had successfully presented the union's position as unreasonable and "featherbedding." The company guarantee which appeared in the newspaper looked impressive. Few people outside of the labour movement recognized the "Catch-22" dilemma underlying the company's promise. If the union gave up the contracting-out clause, the company, as the sole judge of "economic reasons," could lay off workers for economic reasons and contract out much of the bargaining unit work. The union would be destroyed.[19]

The company was aided in making the workers appear unreasonable by sensational press coverage about cable slashing and sabotage. One incident at the beginning of the strike put the union's office telephones out of service for a few hours. Cable slashing is a regular occurrence in the telephone industry, but during a dispute it receives much publicity. In spite of statements

Keeping vigil outside the Boot during the 1977-78 lockout.

Leigh Taylor

condemning such actions, the union is seen as the perpetrator. The union's insistence that it is not responsible and that, indeed, if telephone workers wanted to sabotage the telephone network, their skills and knowledge would allow them to do a far more effective job, does not win the union public support.

There was one instance where strikers in the Lower Mainland came up with an idea which resulted in the laying of charges by the police. The plan was to tie up long distance lines by holding a "phone-in." Strikers were to call directory assistance in Ottawa and talk as long as they could, thereby loading long distance circuits without any cost to the caller. When reporters uncovered this scheme, the workers abandoned the plan. One striker was charged with public mischief, but the proceedings were eventually stayed.[20]

In the second week of the lockout a number of areas began mass picketing. BC Telephone obtained an injunction against mass pickets in Nanaimo, limiting the union to two pickets per entrance. Non-union workers who supported the TWU efforts risked victimization. In Port Moody, a bank worker was fired when she refused to handle customer payments of BC Telephone bills. Her firing brought a group of union pickets to the Port Moody branch of the Bank of Nova Scotia, but the bank refused to rehire the clerk and sought an injunction against the union pickets.

By early December, with the approach of the Christmas holidays, bitterness over the lockout increased. In the coastal areas pickets were drenched with continuous, cold winter rainfalls. In the interior, cold winds and snow made picketing an ordeal. Both company and union officials started to characterize the dispute as "dirty."

On December 5 the union invited the company to submit a joint request for the intervention of the federal labour minister, John Munro. The company maintained that outside intervention was unnecessary and that it was willing to meet face to face with the union. The union went ahead with the request and the minister appointed Michael Collins, the chief conciliation officer in the area, to handle the dispute.

Two days before Christmas the BC Federation of Labour entered the dispute. At a demonstration-cum-Christmas party and carol singing outside the company headquarters, BC Federation of Labour president Len Guy announced a "hot" declaration on all joint Hydro and BC Telephone poles. Cy Stairs, research director of the BC and Yukon Territory Building Trades Council, promised the full support of the construction trades and in particular the IBEW.[21]

Mediation sessions continued on and off through the rest of December with the aid of Collins. They proved fruitless. In early January Munro appointed Supreme Court Justice Henry Hutcheon as industrial inquiry commissioner in the dispute. Announcing Hutcheon's appointment, Munro asked the company, the union and the workers for a return to work pending Hutcheon's report. MacFarlane immediately replied that the workers would be permitted to return to work only if they met three conditions: each employee "must provide his supervisor with a clear-cut, written commitment that there will be no further job action and no further strike action," a guarantee to "work productively," and assurance that there will be no damage to company property.[22]

The union rejected this proposal as a yellow-dog contract. Donnelly explained that "the implications made by Mr. MacFarlane that employees who have served this company faithfully for many years would damage company property is insulting and contemptible." He suggested that MacFarlane had "deliberately created a situation that he knew would leave the union in no position to recommend a return to work."[23]

Eight weeks into the lockout the company claimed that employees had sabotaged equipment before leaving the buildings. MacFarlane explained that the work had been done "with such skill and to such an extent that only the dedicated work of our management people — and a considerable measure of good luck — prevented a major breakdown of our network." Newspaper reporters were sceptical. They asked company officials why it had taken eight weeks to release the information. A company spokesperson replied that the sabotage was internal and did not really affect the public. The reply was unconvincing, especially in view of the company's generally aggressive propaganda campaign. Doug Booth, a vice-president of the union, pointed out that the charges were impossible for the union to address. "How do you prove or disprove what he's saying?" asked Booth. He reiterated that MacFarlane's proposals for a return to work were insulting.[24]

Throughout January the union escalated its picketing, with flying squads closing post offices, provincial government buildings, ferries, car rental firms, shopping centres and construction sites. Wherever BC Telephone supervisors went, flying pickets followed. When supervisors attempted to empty coin phones in shopping centres, pickets were set up. When supervisors boarded ferries, pickets were sent to the ferry depots. A number of stores and banks sported signs provided by the TWU announcing that they would not accept payment of BC Telephone bills until the end of the dispute.

In the second week of January the union started a campaign of mass picketing at key BC Telephone buildings around the province. The campaign ended when the company obtained a province-wide injunction which restricted picketing to two pickets per entrance. Increasingly during January the union found itself in court. Companies struck by flying squads sought injunctions. Five of the cable television companies hit by the BC Federation of Labour's "hot" edict took the federation and the IBEW to court. The courts granted an injunction which ended the federation's use of "hot" edicts over the handling of the joint poles.

The union sent pickets to the homes of supervisors. Stickers began to appear stating, "I scab for BC Tel" and "BC Tel scab works here." Donnelly explained that this new tactic was aimed at "supervisors who don't understand that they are stealing our people's jobs."[25]

On January 25 Hutcheon's report was released to the union and company by labour minister Munro. Hutcheon essentially concurred with Hall's recommendations, with two significant exceptions: the duration of the agreement and contracting out. Hutcheon recommended a three-year contract in place of Hall's one-year agreement. As the union had been without a contract for thirteen months, a one-year agreement would be ludicrous and a two-year agreement would see a return to the bargaining table in six months.

On contracting out, Hutcheon drew up a new method for resolving differences. He took the company's suggestion for a standing committee but increased the terms of reference of the committee and gave it muscle. Hutcheon recommended that the first section in the current contracting-out clause, which specified the 33 jobs which the company was permitted to contract out, be retained. In place of the remainder of the clause, Hutcheon recommended eleven additional clauses which established a contracting out and technological change committee (COTC) comprising four union and four management appointees and an independent chair, appointed by Supreme Court Chief Justice Nathan Nemetz if the parties could not agree on a chair. No contracting out was to take place until the committee or its chair had approved the action. The chair had the power to call meetings of the committee and to decide if the issues brought to the committee by either the company or the union fell properly in the realm of the committee. But by retaining the wording of the former contracting-out clause, the committee was significantly restricted in what contracting out it could permit. Where the COTC failed to come to a decision, the chair was to render a final and binding decision. Finally the clause was to include the guarantee that regular employees with two years or more service would not lose their jobs as a result of technological change.[26]

For more than a week the union and company met with federal mediator Collins. Early in February the union and the company came to a tentative agreement based on the Hutcheon report. The wage settlement was 6 per cent in 1977, 4 per cent in 1978, and 6 per cent in 1979.

The union began a ratification vote on the proposed agreement. In spite of the tentative settlement the situation was still tense. Union officials noted that there was a "huge amount of resentment" against the company. The negotiations were not yet over, since the union and company still had to draw up a return-to-work agreement. "These things are never easy," warned Donnelly. "The longer and more bitter the dispute, the more difficult they become."[27]

The return-to-work negotiations proved to be a major stumbling block. The company insisted that each worker sign an interim contract, which included a statement of the employee's willingness to work alongside management personnel, acceptance of the 1976 rates until the contract was formally signed, and a guarantee that there would be no job action or requests for time off. Workers were

Leigh Taylor

The police were compelled to enter the 1977-78 dispute.

to be called back over a nine-day period. Workers who refused to sign the company's pledge would not be permitted to return to work. Aside from the fact that the terms were unacceptable, the concept of workers signing individual contracts with the company was entirely rejected by the union. The union regarded the company's demands as a contravention of the federal labour code and an attempt to undermine the union's bargaining authority. The TWU was adamant that all members would return to work together.[28]

By Thursday, after four days of negotiations on the back-to-work issue, the union decided on a new tactic. The TWU broke off negotiations with the company. Donnelly anounced that members were returning to work *en masse* on Monday morning whether or not a back-to-work agreement was signed. The company remained unyielding. The prospect of thousands of workers returning to work with or without an agreement and congregating in the hundreds or thousands outside BC Telephone buildings around the province brought the RCMP and city police into the dispute. They pressured the company into eleventh-hour negotiations. After ten hours of continuous negotiations on Sunday, February 12, the contract and return-to-work agreement were finally signed.

More than 60 per cent of the workers returned on Monday, and by Wednesday the entire staff was back at work.[29]

The union and the company issued a joint press release which described the agreement as a "compromise solution" reached "in a spirit of co-operation and an earnest desire to achieve a harmonious resumption of work." With false optimism the release stated that "Both sides are confident that the agreement reached will be the start of a new industrial relations climate between the parties."[30]

After sixteen months in dispute over the agreement, most workers realized that the optimism after the signing of the contract was little more than wishful thinking. The lockout had lasted as long as 95 days in some cases, and 81 days for most TWU members. The central issue in the dispute was the continued existence of the union. Though the contracting-out clause had been modified, it had not been eliminated as the company had proposed. Union members in fact viewed the new clause as an improvement over the previous arrangement.

The union had fought a successful defensive battle for its very existence. But the fight was not over. The fact that the union and company had not even been able to negotiate a back-to-work agreement without the union threatening to go back *en masse* indicated that the settlement was simply a pause for both sides before the next round.

Chapter 10 *The Occupation*

In the wake of the 1977-78 lockout the atmosphere at BC Telephone remained tense. The end to the long, bitter dispute—with the workers threatening to go back to work and the police pressuring the company to settle—almost guaranteed continued antagonism. To top things off, at the end of the year BC Telephone was found by the Anti-Inflation Board to have excess revenues of $25.3 million. The company was allowed to carry this superprofit into its revenues for the following year. When by the end of 1979 it became clear that BC Telephone still had not been able to absorb all the excess revenues, the Canadian Radio-Television and Telecommunications Commission (CRTC)—the new federal regulatory body which replaced the Canadian Transport Commission—ordered the company to refund subscribers $7 million of its inflated profits.[1]

To the workers, the $25 million represented the company's accumulated savings of an average $2,500 in lost wages from each of the 10,000 employees locked out for two and a half months in 1977-78.

In March, 1978, the union held the division council meetings which had been postponed from January because of the lockout. Along with the council meetings the TWU held a special one-day convention to elect a new secretary treasurer and to hold a "post mortem" on the dispute. Don Bremner, a New Westminster craft worker and chair of the plant council, was elected secretary treasurer. Bremner succeeded Tom Stanley, who had died suddenly in November, 1977.[2]

The special convention's discussion on the lockout was lively. Donnelly complained that too many locals had acted on their own, ignoring the union leadership's direction. Strikes require battlefront discipline, he explained. Once the union developed a strike strategy, it was vital that all locals follow the planned action.

Convention delegates objected that they had received little information in the field during the dispute. Many areas learned of developments through newspaper reports. Local strike leaders also resented being told to remove pickets or stop actions because the leadership felt these jeopardized the legal or strategic position of the union. Underlying the complaints of both the councillors and the leadership was frustration with their inability to prevent the company from locking them out or to generate pressure for a quick settlement once the lockout was in effect.

Review and analysis of the lockout continued at the summer convention. The long job action had once again spurred the union into constitutional change. The 1976-77 constitution, while clearly a move towards consolidation, did not resolve the continuing problems in communications and internal unity.

A special joint council meeting of the three divisions was held in January, 1979. This meeting struck a number of committees, each charged with making recommendations for changes in the union structure and with preparing for a possible dispute over the 1980 contract. The committee empowered to make recommendations on union structure was aptly named the "one-union concept committee." A tactical committee was authorized to "draft contingency plans that will be implemented in the event of another dispute with the telephone company." A further committee, the strike structure committee, was charged with the tasks of identifying problems that occurred during the lockout and recommending improved structures for use in future disputes.

The summer convention had a heavy agenda. Held four months before the TWU and the company opened bargaining, the five-day convention discussed a report from the bargaining committee and established bargaining priorities, adopted a totally new structure for the TWU which abolished the divisions, worked out organizational structures for strike coordination, and began a discussion on possible job action strategies in the event that the company forced another confrontation.

The one-union concept committee reported that, in spite of the 1976 constitutional changes, there remained major structural problems. For the most part, these problems could be traced to the duplication of effort and dispersion of resources caused by the divisional structure. On the basis of this report a new structure was adopted by the convention and submitted to the membership for ratification.

Under the new structure the executive was to consist of all salaried officers of the union—the president, the secretary treasurer, a single vice-president, and twelve business agents. All locals were to become locals of the TWU, and the power to charter new locals would rest with the executive. Business agents could be assigned to any local. In place of the annual winter council meetings and summer convention, the new constitution established two week-long conventions to be held in January and June. The convention would determine the bargaining goals of the union.[3]

The strike structure committee designated three major problem areas—internal communication breakdown, lack of liaison with the leadership, and a wide variance in strike policy from area to area. The committee recommended a tight strike administration with a central strike director and a communications officer to be appointed by the union executive no later than September and zone coordinators to be appointed in the fall. The strategy was to have local strike committees in place long before action was necessary.

The tactical committee opened its report by explaining the TWU's need for such an extraordinary committee. The committee noted that the "normal collective bargaining process does not function for the TWU in that the company has no economic incentive to bargain in good faith . . . This forces us to adopt unorthodox methods to further our interest." The tactical committee recommended that "the executive council implement a policy of preventing the Company from forcing us off the job and that this union adopt as our initial strategy a policy of bringing economic pressure to bear on the company while remaining on the job as long as this policy is effective." Much of the discussion focussed on how the union could achieve these goals. The union was acutely aware of its failure to gain public support in the last dispute and the resulting public image problems which plagued the telephone workers through the entire lockout.[4]

In spite of the adoption of the new union structure, many of the bargaining goals of the union were worked out before the convention by the three divisions. This convention still adopted only the general priorities for the master agreement while the divisions developed the goals of their separate appendices. The priorities for the master agreement were an across-the-board wage increase, improvements in vacations and accumulated time off, an increase in the company contribution to the pensions, and a general increase in the employees' benefits package. Among the non-cost items the number one issue was job security. As well, the union sought the merger of the company pension plan with the jointly trusteed bargaining unit employees' pension plan. In 1973, when the union won the right to a jointly-trusteed pension plan, the company separated the employees' pension plan—for bargaining unit employees—from the company's own pension plan for supervisors and non-bargaining unit employees.

Talks on the 1980 agreement began in October, 1979. After two and a half months of little progress, the union demanded that the company submit a wage proposal. The company complied in January, but insisted that its wage offer of 7 per cent in 1980 and 8 per cent in 1981 was a total cost package. As the company had just granted supervisors salary increases of 10 per cent and 14 per cent,

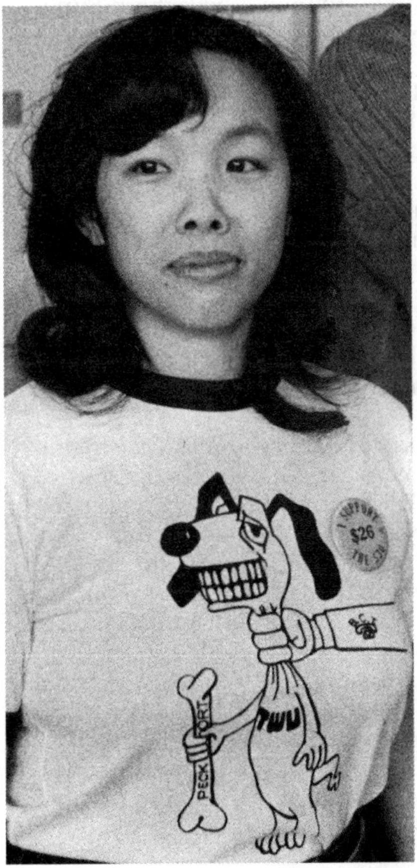

Lila Wing's T-shirt triggered the takeover.

this offer totally outraged the union. By the end of January conciliation officers Michael Collins and Archie McDonald had been assigned to the dispute. Once again the union and the company had failed to reach agreement before the old contract had elapsed.

The union realized that it faced a long dispute, so the TWU initiated the "Super Service" campaign, aimed at pressuring the company to settle. Super Service provided special service to the customer, with telephone workers taking greater care in servicing customers. Production sunk to an all-time low. It was a type of work-to-rule campaign, where workers follow company regulations to the letter. BC Telephone, like most large companies, has hundreds of rules which are not regularly followed in complete detail. For example, installer and repair workers preparing to leave the compound each morning were required to make a thorough check of their vehicle, including testing the horn, brakes, headlights, turn signals, tire pressure and four-way flashers. Super Service amounted to using the company's own rules to bring about a work slowdown.

At the end of February Ed Peck was appointed conciliation commissioner. Peck met separately with the parties for four and a half months. Throughout the conciliation process, the pressure grew. In March the company announced that it was seeking a rate increase from the CRTC. It wanted permission to raise rates 12.5 per cent for residential service and 15 per cent for business lines effective December, 1980. It further requested that it receive a portion of the increase as an interim hike effective June 1. Coincidentally the amount requested in the interim increase worked out to be $7.8 million, almost the same amount that the company had three months earlier been forced to refund to subscribers.[5]

In May the CRTC rejected the company's application for an interim increase and set September 30 as the date for the start of public hearings over the full application. The company's reaction to the refusal of an interim increase was to threaten layoffs. The union condemned the layoff threat as an attempt "to use the workers as pawns to exert pressure on the CRTC." The union also announced that it would intervene in the upcoming CRTC hearings on the issue of quality of service.[6]

On August 27 Peck completed his 88-page report, which the union and company received early in September. Peck opened his report with a 9-page introduction which described the "tortured" history of collective bargaining between BC Telephone and the TWU. He noted quite simply that the collective bargaining system "is not working as intended" at BC Telephone. Peck described the relationship between company and union "as highly adversarial, with destructive overtones." He explained that part of the problem was attitudinal, with the union "deeply suspicious, by its own admission almost 'paranoid' in its concern over company objectives." Peck observed that the company was "reluctant to accept that the union has a responsibility to its membership in respect to plans and changes which impact on staff." He characterized the company as having "an exaggerated notion of its management rights."[7]

The Peck report contained 73 recommendations for inclusion in a proposed two-year agreement. These included wage increases of 75 cents an hour across the board effective January 1, 1980, an additional 5 per cent effective July 1, and a 10 per cent increase effective July, 1981, six additional days accumulated time off, improvements in vacation benefits, and changes in grievance and arbitration procedures. Peck also made a number of recommendations on the highly contentious issues of work scheduling and jurisdiction.

The union's bargaining committee, headed by recently elected TWU president Bill Clark, voted immediately to recommend acceptance of the Peck report to the membership. Membership meetings were held across the province. Before the union could start the vote, however, the company announced that it had rejected Peck's four key recommendations concerning wages, accumulated time off, scheduling and jurisdiction. The union nevertheless proceeded with the membership vote. Clark explained "by no stretch of the imagination did the Peck report come anywhere near to what we had demanded of the company, but we understand that in the process of negotiation there has to be give and take . . . We did not expect to get everything and we believe the company should not get everything they want, obviously they thought they should."[8]

BC Telephone negotiators met with the union and tried to persuade the union to vote on the company's final offer in place of the Peck report. But the bargaining team rejected the request. Although the company accepted 69 out of 73 proposals in the report, the four that it rejected were deemed crucial by the union.

The company had informed Peck during conciliation that it could not afford a wage increase greater than its 7 per cent and 8 per cent final offer. Peck nevertheless had recommended more, based on recent wage settlements in British Columbia and the fact that the TWU had lost ground on wages under the AIB controls program. The company rejected Peck's wage recommendations on two grounds. It argued that the amount was "in excess of the company's ability to pay" and that the 75 cents across the board increase "would cause greater compression between top bargain-

. . . but only temporarily.

ing unit rates and other bargaining unit rates." The company was adamantly opposed to union demands for wage compression as it made "higher jobs groupings less attractive." The company rejected Peck's recommendation on accumulated time off because it would be too expensive.[9]

On the two non-cost items—jurisdiction and scheduling—BC Telephone argued that Peck's suggestions challenged management's rights. The company rejected Peck's recommendation that the union and company write job descriptions for plant classifications, with either party having the option of sending a disputed description to arbitration. Management argued that written job descriptions in plant would "seriously affect the company's ability to make necessary operating changes." Similarly it rejected Peck's recommendation that on assignment of shifts or out-of-town work, management determine the number of employees required and the union schedule committee assign the employees.

By mid-September the TWU membership ratified Peck's report with a resounding 91 per cent in favor. With their members solidly behind them, the leadership turned to the selective strike. They wanted to avoid a long-drawn-out battle of attrition, with the union holding rotating strikes and the company responding with lockouts. President Bill Clark outlined the union's proposed strategy to about 500 construction, PBX repair workers and installers, and special services workers. The workers, asked if they would be the front line of the union's selective strike, overwhelmingly accepted the assignment.

On Monday, September 22, the union began its new job action, with 530 craft workers reporting to work and refusing all assignments except emergency repair work. The workers had been chosen because they worked in the company's most profitable areas. Special services workers were responsible for data and computer services; the construction group did the prewiring of communication lines in new buildings; PBX installers and repair crews worked on large business switchboards. The selective strike was aimed at BC Telephone's money-making areas, its major business accounts; it did not affect the vast majority of telephone subscribers. The company also would not benefit from the saved wages of a lockout as most employees were still at work.

The striking workers reported to work and then "sat in" in the coffee rooms, garages, or spare rooms in the compounds. They played cards, chess, checkers . . . monopoly, some even played basketball and volley ball in the parking lots. There were no pickets because the strikers did not want to force the remaining employees off the job. As the striking workers were carrying the battle for the whole union, they were paid 70 per cent of their gross wage from the union's strike fund. To help fund the selective strike, the 10,500 employees remaining at work were asked to contribute thirteen dollars a week to the strike fund. Clark felt that the selective strike would place maximum pressure on the company "without picketing all over British Columbia."

Within weeks the selective strike produced a significant backlog in construction and switchboard installation and repair. The company began sending out supervisors to replace the striking employees. The union responded by following the supervisors to the job sites with flying picket squads. As supervisors left the BC Telephone buildings, they were followed through the streets by union pickets. Some supervisors, not overly pleased with their role and recognizing that eventually they would have to work with the striking employees again, informed their flying squad tails at each stop where their next stop would be. Often the supervisor would wait while the pickets warned the customer that the company service they had requested was strikebound and pickets would be following the supervisors. Most companies with an organized work force decided to wait until the end of the dispute rather than risk a picket and a shutdown of their job site. In spite of the civilized and low-key manner in which much of this picketing took place, it was frustrating. The supervisors were doing the employees' work; each job completed meant that much less pressure on the company. Adding to the frustration and anger felt by workers following the supervisors and observing them while they did bargaining unit work was the fact that a minority of supervisors decided that it was their job to attempt to lose union pickets by racing through yellow lights, driving the wrong way up one way streets, or pulling into large parking lots and pulling out by a different exit.

Early in October negotiations reopened with mediator Michael Collins, but talks broke down. Two linemen in Prince George who were off the job as part of the selective strike were so frustrated with the failure of talks that they decided to force a settlement on their own. Late on Sunday, October 19, the two workers, Peter Massey and Don Gordon, entered the Prince George BC Telephone office and barricaded themselves inside two rooms containing electronic telephone switching equipment. The pair demanded that the company sign a contract or they would cut off the city's commercial telephone service, leaving only residential and essential service. They warned the company that they had booby-trapped the computer and that if the company or police attempted to force their way into the rooms, they would wreck the computer equipment by spilling water on it.

The union found out about the occupation early Monday morning, when the two workers telephoned the union office. Bill Clark talked with the strikers, promising that the union would do what it could to get them out, but that they had overstepped the bounds of industrial relations. On the urging of Clark, Jack Webster, a popular television open-line broadcaster, negotiated an amnesty for the two strikers by mediating over the telephone with both the union president and the BC Telephone chairman Gordon MacFarlane. Webster elicited from MacFarlane the promise that no legal or disciplinary action would be taken against the two provided they did no damage and left company property immediately. Clark persuaded the workers to accept the amnesty and leave the building for the duration of the dispute.

Though successfully resolved, the incident illustrated the bitterness developing in the dispute. Massey explained that the occupation was designed to deflect their co-workers' frustration and so avoid destruction and violence. Both Massey and Gordon had felt that their action could resolve the dispute and take some of the heat out of the growing confrontation. "It's like lancing a boil to take a little pressure off," explained Massey.[10]

Through October and November the pressure continued to build with attention focussing on the CRTC rate increase hearings which opened on September 30. The union had decided to oppose the company's application on the grounds that any increase should be contingent on an improvement in service. In April the CRTC had held brief quality-of-service hearings where the union made proposals on the need for the commission to develop its own criteria for judging quality of service, independent of those offered by BC Telephone. The union pointed out that BC Telephone relied heavily upon a customer survey purchased from its parent, GTE, which did not adequately measure customer dissatisfaction with company service. Key indicators—for example, the amount of time a customer must wait to have a new telephone installed—were not included. Nor did it correlate the time a customer must wait for help from an operator with the number of operators that the company had allotted for a particular shift. Some of the most frequent customer complaints, such as noise on the line or having to continually re-dial, were not part of the company's in-house survey.

Prior to the start of the hearings, the union called for the CRTC commissioners to hold regional hearings in outlying areas. Because BC Telephone planned to centralize its operations by introducing electronic switching that would allow the company to close traffic centres in Vernon, Nelson, Dawson Creek, Terrace, Cranbrook and Williams Lake, the union regarded regional hearings in these centres as crucial. The TWU estimated that the elimination of these traffic centres would remove 850 jobs from the communities and millions of dollars from the local economies. Working closely with the city councils in these areas, union business agent Gordon Stoutenberg argued that "this is not just a union issue, the public has to be made aware that this move by BC Tel will have a dramatic effect on the communities involved."

The hearings were the longest in Canadian history, lasting 43 days. Before the commissioners, the TWU argued that the company's massive automation campaign was not designed to improve service but rather to create an outlet for purchasing GTE equipment, with telephone subscribers in British Columbia footing the bill through higher rates. The union explained:

> It is important to note that GTE is a massive, vertically integrated multi-national corporation which owns a range of telecommunications companies. A major portion of GTE's revenues is derived from the manufacture and sale of telecommunications equipment. GTE's subsidiaries are its most dependable customers. They function as a built-in market for equipment produced by the parent company and its other subsidiaries. Unlike independent companies that buy equipment only as it is needed, BC Tel's purchasing and construction programs must conform to its parent company's desire to sell.
>
> At a time when many corporations are being forced to curtail their operations, telecommunications manufacturers and operating com-

Supervisors leave the BC Tel building in Nanaimo after workers began their occupation.

André Bérubé

panies are in the process of a massive expansion program. BC Tel's plans for record levels of capital expenditures fit in with a general expansion planned by telephone companies throughout North America.[11]

Union witnesses testified time and again that the company was reducing the quality of service to customers, while at the same time driving up the rates of telephone service. Bob D'Etcheverrey, an installer-repair worker with BC Telephone for sixteen and a half years, explained how the company "is conditioning customers to accept worse service and to do more and more of the work themselves." D'Etcheverrey used the example of BC Telephone's Phone Mart system, designed so that the customer does most of the work, but is still billed for service.[12]

Jim Gordon, a craft worker with BC Telephone for 23 years, testified:

> BC Tel talks a great deal about the need for improved quality of service. But, whenever there is a budget cut in the company, the first thing that is cut is transmission improvement and preventive maintenance. The manager who is in charge of budgeting and budget cuts is not the same person who has the direct personal contact with subscribers. So when the meat axe comes down, he hacks the easiest thing to cut. Unfortunately, this means that our

customers suffer. Customers are paying for the ability to speak and hear clearly. But the company does not make it a priority to provide high quality transmission or to reduce the level of noise that is on the lines. There are approximately 1,000 installation and repair men in the company. But only 400 of these men are provided with noise and level meters which are the key tool needed to diagnose the source of noise problems. With such instruments, testing for noise requires as little as 5 to 10 minutes of a repairman's time. But, without these meters, the task of locating the source of noise becomes infinitely more difficult and time consuming — if it is done at all . . .

It appears to me that the Company has shifted its emphasis. We used to provide high quality service. That was at the heart of everything we did. But, in the last several years, service has become a secondary priority. New equipment is slapped into place and left to deteriorate. When the equipment fails, a quickie patch-up job is done or new equipment is installed. The result of this change in policy is that the customer suffers inferior service and increased down time.[13]

Russ McNicol, an installer and repair dispatcher with BC Telephone, explained how the company's measure of an installer's productivity, known as RE (reaonsable expectancy), has been increasing over the years, resulting in speed-ups and declining quality of service:

> Workers are disgruntled. There is no longer the sense of pride that used to exist among employees at BC Tel. More is being demanded in terms of quantity. The result is a decline in the quality of work. In order to satisfy productivity requirements, quality of work must suffer. When I was an installer, it was more or less left up to me to take the time necessary to do a good installation job. Now, however, an installer will be given a job which is rated with a certain number of REs, he has to rush the job. It's a Catch-22 situation—either do a sloppy job and keep your supervisor off your back or do a good job for the customer and risk getting upbraided by your supervisor.[14]

The union members contributed testimony that no other group could—an insider's view on the operation of the company. In the past, hearings had been little more than forums for management experts who used mountains of material to defend their rate requests. In the 1980 rate hearings, however, BC Telephone workers helped demystify company evidence. While the other intervenors opposing the company's claim—including the National Anti-poverty Association, the Consumer's Association of Canada, the North Delta Ratepayers Association, the Department of the Attorney General, the BC Federation of Labour, the International Woodworkers of America—were outsiders looking in, the TWU members saw from the inside the flaws in the company's testimony. The telephone workers argued from experience that capital expenditures should not necessarily be equated with improvements in service.

In its innovative submission the union argued that "before any rate increase is granted, the Company should be obliged to file a second application with the Commission . . . demonstrating that any increase in revenues will be spent in improving the quality of telephone service." The union noted that when Chrysler Corporation requested financial assistance from the governments of Canada and Ontario, the company had been required to meet specified conditions before the loans were granted. The TWU suggested that "the CRTC should use this same approach to ensure that BC Tel will behave in a socially responsible manner."[15]

The hearing placed BC Telephone on the defensive. The message sent out by the union to the public was a loud and clear "don't blame the workers for the mess at BC Telephone." The union had long been trying to get this message across, but the CRTC hearings provided the first real breakthrough. As TWU business agent Theresa North explained, "TWU members cannot accept criticism that portrays workers and management as equally responsible for the poor quality of phone service. Consumers deserve the finest quality of phone service that can be delivered. The TWU would like to be able to deliver such service. It is BC Tel that frustrates the desires of both consumers and workers."[16]

A further issue which the union raised during the hearings was the demand for nationalization of BC Telephone. Recognizing that the CRTC did not have the power to nationalize BC Telephone, the union nevertheless contended that this was the ultimate solution to the problems of quality of service and ineffective regulation of the foreign-owned company. Towards the end of the hearings, the union began a public campaign in which it sought signatures for a petition calling on the government of British Columbia to make BC Telephone a crown corporation. The union argued that as long as the telephone company was in the hands of a multinational, it would not serve the needs of the people of British Columbia. Since the company was a utility, the TWU said, it should be publicly owned and regulated.

The union was confident that the company would not attempt a full-scale lockout while the rate increase hearings were in progress. In November the company obtained an injunction against the union's flying pickets, limiting picketing to two per building entrance. On December 1, three days before the hearings ended, the company was granted a variance in the original injunction. The new wording of the court order allowed the company to expel the

530 sit-in strikers from company property throughout the province. It restricted the union

> from interfering with the employees of the Plaintiff by preventing or attempting to prevent by the use of force, threats, intimidation or coercion the said employees from entering, leaving or carrying on work in any of the places of business of the Plaintiff in the Province of British Columbia and *from trespassing on any premises owned, leased or otherwise in the possession of the Plaintiff in the province of British Columbia by sitting in and refusing to leave such premises within 10 minutes of being told by the Plaintiff to leave and not return until notified by the Plaintiff.* (author's emphasis)[17]

In mid-December the company began disciplinary action against employees who visibly supported the "Crown Corporation Now" campaign promoted by the TWU. Employees were warned that suspensions would be handed out if they wore the "Crown Corporation Now" button. Management personnel removed the nationalization petition and information from union notice boards. Employees were told that their vehicles would not be permitted in company parking lots if they had "Crown Corporation Now" stickers on them.

On Friday, December 12, over 1,000 employees were sent home from work for refusing a company demand to remove the offending buttons. The TWU secretary treasurer, Don Bremner, condemned the company's suspensions as "an attempt to escalate the present dispute." Bremner commented, "It is extremely difficult to believe that BC Tel would subject the citizens of BC to a full-scale labour dispute because they wish to deny their employees the right to freedom of expression." Faced with an immediate mass lockout over the button, the union leaders decided "to avoid the escalation that this company so obviously wants," and asked their members to refrain from wearing the buttons until the issue was resolved in arbitration.[18]

Early 1981, negotiations started again with the aid of federal mediator Michael Collins, but broke off by the middle of the month. A week later the company initiated a campaign of selective suspensions. BC Telephone, taking a leaf out of the union's book, decided to soften the union through escalating suspensions. Starting in mid-January, the company suspended a few hundred workers a week. The union, having paid the original 530 strikers 70 per cent of their gross wage, continued the policy for the additional workers put on the streets through the escalating suspensions. The selective strike alone was taking a toll on the union. Even with the thirteen dollars a week paid by employees still at work, the strike fund was becoming depleted. The company recognized that eventually the growing numbers of locked out employees would drain the union's dwindling strike fund, leaving the TWU with all of its members locked out and no money. The company also avoided possible unfavorable publicity which would accompany a mass lockout.

On Monday, January 19, BC Telephone suspended 220 installers "for low productivity." Company spokesperson Keith Matthews denied TWU charges that the action was a lockout. Matthews said that the company considered the installers to be on strike. He assured reporters that the company had "no plans for a lockout." But when the union advised the installers to report for work, the company ordered them home. Clark termed the company's move a "sophisticated way of starting a lockout . . . If the installers are not working, then the clerical and support staff will have nothing to do." He said the company would then accuse these workers of low productivity and send them home. A week later, a further 85 workers were laid off for lack of work. By the end of the month nearly 1,000 members were off the job.[19]

On January 29 the CRTC brought down its decision on the company's request for a rate increase. BC Telephone was granted everything it had asked for: a 12.5 per cent increase for residential service and 15 per cent for business service. The CRTC predicated its approval of the increase with a warning that a "minimum acceptable level of service quality" had to be reached by the end of 1981 or the commission would take "action appropriate to the response of the company."[20]

Clark observed that "the commission granted the rate increase before BC Tel had brought the quality of service up to the level that telephone customers have the right to expect." While the union was pleased with the commission's emphasis on quality of service, Clark suggested that the rate increase "should have been done on an incremental basis in the nature of a performance bond."[21]

An operator enjoys workers' control in Nanaimo.

André Bérubé

The membership of the TWU believed that now that the company had received its increase from the CRTC, a total lockout was imminent. The union's strategy, unchanged since September, was faltering in the face of the company's selective lockouts. Since December the union had been on the defensive, reacting to initiatives taken by the company. In closed sessions at the convention in January the union had discussed various actions which could be taken in the event that BC Telephone opted for a general lockout. The occupation of BC Telephone buildings was suggested. Larry Armstrong, the union's strike coordinator, asked local strike captains to discreetly poll their members and ascertain whether they would be willing to stay on the job in case the company attempted a full-scale lockout.

Shortly before quitting time on Tuesday, February 3, 21 maintenance workers in Nanaimo and 8 in Duncan were suspended for "going slow." The workers interpreted the suspensions as a prelude to a full lockout in Nanaimo. The workers gathered in the lunchroom of the company's headquarters on Fitzwilliam Street and after consultation with the TWU office in Burnaby, they occupied the building.

With this action the union quickly recaptured the initiative. The occupiers secured the doors and posted union members at the main entrance. The door committee demanded a show of union cards and checked the identification of people who sought admission to the building. Management personnel were allowed to remain in the building, but they were relegated to a suite of offices on the ground floor. As workers replaced all supervisors, they took responsibility for the staffing of boards and maintenance of switching equipment. From late afternoon until well into the evening, shop stewards contacted workers and set up shift schedules to cover the boards 24 hours a day and provide security for the buildings.

By evening reinforcements began to arrive with sleeping bags, snacks and provisions for a long stay. The unionists vowed to stay in the buildings "until we get our contract." Alf McGuire, a TWU councillor in Local 3 and one of the suspended employees, told reporters, "we're just your common or garden switchmen, and when people, ordinary people, get desperate enough to take a building over things are getting pretty desperate." In reply to the company's claim that the switchmen had not been producing, McGuire explained, "when you've been sixteen months without a contract you're not exactly a star performer . . . Morale has been very low and hasn't been getting any better." McGuire charged that the company had been keeping "everyone in a state of turmoil and upset" which left people "more and more frustrated."[22]

The occupation brought about a complete change in atmosphere in the Nanaimo telephone building. Grinning workers and people enjoying their jobs could be seen everywhere. A makeshift banner announcing "Under New Management, TWU" was hung from the microwave tower on top of the building. Smaller door signs proclaimed, "BC Tel, Now 100% Canadian Owned." The union encouraged operators to give the best service. Operating boards were fully staffed. Clerical and craft workers were taught the rudiments of operating by experienced operators and many worked full shifts on the boards. For once the boards were fully staffed. "It's almost been a carnival since we took over," commented McGuire, "people are glad to be free of supervisors." As word of the occupation spread throughout the province, BC Telephone workers called the Nanaimo office with messages of support and encouragement. One BC Telephone office took up a collection and sent the money to Nanaimo so that the occupiers could send out for pizzas.

In spite of the jovial atmosphere in Nanaimo, there was an increase in tension in every other centre in the province. The union was in a serious situation. More than two months earlier the company had been granted an injunction which specifically prohibited sit-ins. While the union saw its action as a defensive move aimed at preventing a company lockout, few thought that the courts would side with the union. The TWU executive met all day on Wednesday, February 4, to discuss a course of action. Two union officers went to Nanaimo to view the occupation first-hand and report back to the executive.

In telephone offices throughout the province workers spent February 4 discussing the Nanaimo action. Workers asked themselves whether they would participate in a similar action. Nanaimo provided a valuable example. Newspaper articles and television news items revealed that the occupation was peaceful and the workers were enjoying themselves. Initial reaction of the public and press was not unfavorable. The news reports clearly showed that the action was not a desperate act of an isolated minority. Allan Stratton, a TWU member in Nanaimo, expressed a widely held view when he stated, "We're not playing snakes and ladders here. I've got a wife and kids. I need to get a decent living out of this company and I'm going to put my job on the line for it."[23]

On Thursday morning the TWU extended the occupation to the rest of the province. The incident which sparked the takeover was a supervisor's demand that Lila Wing, a shop steward at BC Telephone's supply depot in Burnaby, go home and change her T-shirt. Wing was wearing a cartoon T-shirt that union members had printed. Other TWU members had been wearing shirts similar to Wing's for months. The cartoon on the T-shirt depicted Snoopy, of Peanuts cartoon fame, wearing a TWU crest and holding a bare bone labelled "Peck report" while being choked by a hand with "BC Tel" written on the cuff. Refusing to leave the building, Wing contacted the union office for direction. The TWU executive authorized the workers to take over the building. At the same time union leaders called telephone workers in fourteen major exchanges in the province with instructions to take over. By noon the occupation had swept the province.

In a press conference at noon TWU President Bill Clark defended the union's actions. Clark explained that the attempt to send home a shop steward because of her T-shirt was the final straw.

> Arbitrary moves by BC Tel Management, telling employees what they can and cannot wear, and many other cases, such as the company's suspension of employees wearing the Crown Corporation button, or the threats against employees who have bumper stickers on their private vehicles, only demonstrates the aggressive, provocative stance of BC Tel management.
>
> The union's worst fears have now been met. BC Tel has received all of their rate increase from the CRTC and are now embarking on a Union breaking campaign, throwing into question not only the right to collective bargaining, but civil issues such as the right to freedom of speech.
>
> In response to the latest provocation by BC Tel the TWU has decided to maintain telephone service in spite of BC Tel's attempt to force a lockout. Our members will be staying on the job and providing basic telephone service. We will be staffing the service centres around the clock.
>
> We are asking the public to support us by refusing to pay their phone bills. Our members will be working without pay and do not see why the company should be collecting from the public for a service we will be providing for free.[24]

Following the pattern established in Nanaimo, the occupied exchanges across the province were soon transformed. Supervisors were asked to either leave the buildings or to remain in designated areas. Most management personnel opted to go home. Supervisors, police or reporters who wanted to inspect the buildings were granted entry and accompanied on their tours by union members.

There were two main assignments in each building: door committees to secure entrances in order to restrict access to the buildings and ensure union control; and committees which supervised the staffing of the operating boards. Strike leaders in some areas spent the full five days in the buildings, but most workers

West Ender

This cartoon by Leo Burdak appeared in Vancouver's *West Ender*.

came and left the buildings according to schedules drawn up by union committees. Workers held meetings in most occupied buildings and worked out shifts, assignments and "occupation rules." In Nanaimo, for example, the locally adopted rules read as follows:

Door Committee:
— only 1 open entrance
— make sure fire exits are accessible
— all other entrances manned
— all windows on ground floor locked
General Rules:
— definitely no equipment damage
— encourage operators to give "best" service
— no alcohol on premises or intoxicated people to come in (put them on another shift)
— set up shift schedules to cover 24 hours a day
— shifts to be 12 hours
— maintain all services, billing tapes, etc.

— keep premises as neat as possible (delegate housekeeping crews)
— if RCMP arrive to expel workers, no force, but don't co-operate (make yourself awkward as possible).[25]

The union executive set general rules of conduct in the occupied buildings. Clark explained that "there is to be no damage and no violence." The union's position in the event of a police attempt to expel the occupiers was to urge members to resist passively by sitting or lying down, forcing the police to remove each worker bodily.

With the workers in control, the regimentation which the company had demanded in the workplace was abandoned. Operators were no longer required to put up a flag when going to the washroom. Breaks were taken when required, and no one was reprimanded for talking too long with a subscriber. If workers found that calls were building up, they recruited operator volunteers from door committees or other workers in the buildings and trained them on the operating equipment. Operators varied their responses from the rigid mechanical replies demanded by the company. In some areas operators decided to answer directory assistance inquiries with "TWU directory assistance" or "BC Tel, under workers' control." Workers rotated their jobs to reduce the monotony. Many workers took tours of the buildings and were shown jobs and tasks which they had only heard about. For many it was the first time they had seen other areas of buildings they had worked in for years.

But the key difference was the atmosphere of cooperation and responsibility. Craft and clerical workers gained new respect and sympathy for operators after experiencing the stress of the operator's job. More than one craft worker abandoned operating after only a few hours, finding it difficult to believe that anyone could work under such conditions for seven hours a day. For the first time in many years telephone workers felt proud of the work they did. The control they exerted over their jobs was limited because the structure of work was still dictated by the machinery. But they were still able to exert some control over their work for the first time since coming to BC Telephone. Most felt a tremendous relief from the knowledge that they were no longer being constantly monitored.

The atmosphere during the five days of the union occupation was unreal. In most areas of the world the seizure of the telephone exchanges by workers constitutes the first act of a revolution. While BC Telephone characterized the occupation as "anarchy," it was generally seen as a further escalation of the drawn-out labour dispute. By seizing the telephone buildings, the union had gone beyond the normal bounds of collective bargaining, but union members felt that their inability to have any effect on the company through traditional tactics made the occupation necessary.

Typical of the publicity around the occupation was an editorial in the Nanaimo *Daily Times* of February 5 which commended the union for its "diplomatic coup" in occupying the exchanges:

> What happens when an unstoppable object hits an unbreakable wall?
>
> The answer takes form today in the current occupation of the BC Tel offices by union staff. The sit-in, camp-in, work-in is seen by both management and union as being the inevitable result of continually worsening industrial relations, and it's a wonder to us that both sides are able to remain as amicable, and respectful of each other as they are on Fitzwilliam Street.
>
> Both sides have said, "what could we do? Our hand was forced!" and the conclusion we come to is if the whole BC Tel-TWU contract issue had been a local one, it would have been resolved long ago. Both management and staff have held together an extremely difficult situation with more than 300 people working without a new contract for more than a year, and their wages tied to figures which inflation has made obsolete long ago. They've seen their management supervisors get two pay raises since their contract elapsed, and naturally they're angry. Just as naturally they have lost respect for BC Tel and who could blame them. After the flip-flop Gordon MacFarlane did after the Peck Report he can hardly be seen to have the best interests of the staff truly at heart.
>
> The natural consequence was an "involuntary, unconscious" go-slow by maintenance workers, and management, obviously, was instructed to get tough.
>
> We have to say that the union has adapted to the situation without losing a single point. Suspended workers have stayed on without pay to keep the equipment working, staff are policing themselves and keeping security tight. They've acted responsibly and should get credit for pulling off a diplomatic coup.
>
> However, it can only last as long as the management lets it, and the way we see it the right answer is to get a contract signed.

A unique set of circumstances allowed the union to win the public's sympathy in this dispute. The long CRTC rate hearings had brought BC Telephone under public fire through most of 1980, with first the interim increase request, then the quality of service hearings, and finally the full hearings from September to December. British Columbians were constantly reminded of the company's large profits. The intervention of the TWU together

Pacific Tribune

TWU president Bill Clark at a mass rally across from BC Tel's Seymour Street building after the occupation ended.

with other labour and consumer groups arguing that a rate increase was not merited and criticizing the company's quality of service left little public sympathy for the company. After the company's refusal to sign the Peck report — in spite of members of parliament, BC Telephone supervisors, and the federal labour minister all urging the company to do so—BC Telephone was condemned for the deteriorating labour relations.

The union's occupation of the exchanges was a peaceful and disciplined action. Its decision to continue basic telephone service meant that the general public was unaffected by the occupation. By continuing to staff operating services, the union was able to show the public, more clearly than any press statement could, its desire to maintain service. In the past telephone workers on the street, whether by strike or by lockout, bore full blame for the decline in service and for the dispute itself. Now, with the union in control and the telephone workers providing service without pay, the company was seen as the belligerent party.

Aside from public support, the situation at BC Telephone had been closely monitored by the BC Federation of Labour. The federation and the International Woodworkers of America had intervened along with the TWU in opposition to the rate increase. At the federation's annual convention in November a special lunch hour demonstration by delegates was organized at BC Telephone headquarters in Burnaby. After the province-wide takeover of telephone buildings, the federation called a special meeting of all affiliates' staff where a strategy was drawn up to support the telephone workers and the three other unions facing intransigent employers at the time. Federation president Jim Kinnaird described the disputes as part of a wider campaign by the British Columbia Employers' Council to "stonewall" on collective bargaining, using the courts, injunctions and industrial inquiry commissions to drag out disputes and avoid bargaining. Kinnaird attacked those employers who abused "the Labour Relations Board, the courts, used strikebreakers and any other method to create the industrial relations breakdown we are presently witnessing." In response to this employer offensive, Kinnaird announced that the federation would be initiating an "escalating program of economic action . . . As far as we are concerned the 'gloves are off' . . . we will win these strikes using the full force of our militant tradition." Kinnaird declared "industrial relations war on the employers of British Columbia."[26]

The following day leaders of the federation, among them Kinnaird and the International Woodworkers of America regional vice president, Jack Munro, underlined their support of the TWU by touring the occupied William Farrell building at 768 Seymour Street. The visit boosted the morale of the telephone workers by showing concretely that the BC Federation of Labour was in the battle with them. The *Province* described the presence of Kinnaird and Munro as an endorsement of the takeover, which constituted a "recipe for anarchy." "Now that the precedent of supporting a takeover of property has been set," asked the *Province*, "in the future might we not expect to see, for example, longshoremen taking over the wharves? Bus drivers seizing their buses? Tellers taking over banks? All could be equally justified."

The occupation occurred at a fortuitous time. Throughout the occupation the press carried stories on the occupation of factories and worksites in Poland by the Solidarity trade union. Most Western leaders were publicly defending the Polish workers and condemning the Polish government. This stance in effect gave legitimacy to occupations as a form of popular dissent. The *Province's* editorialist failed to see the irony in praising Solidarity while condemning the TWU's action.

The openness with which the union welcomed reporters into the occupied buildings made it clear that the workers had nothing to hide and that they did not fear public scrutiny. The press tours also allowed the union to refute company allegations that damage was being done to equipment. As well, the tours gave occupiers an opportunity to make their case to the press first-hand and explain their many long-standing grievances with BC Telephone.

Finally, the vulnerability of the equipment in the occupied buildings made it highly unlikely that the police would risk a surprise expulsion or raid. In fact, as far as the police were concerned, until the courts presented them with a warrant, the occupation was part of a labour dispute and they had no plans to intervene. For its part, the union guaranteed that there would be no damage to equipment. But the situation could change quickly with an attempted expulsion of the workers. While the union had asked workers to resist passively an expulsion attempt, in the heat of such a confrontation it would have been difficult to predict the reaction of the workers or the police. Also any action taking place in one part of the province would have been communicated instantly to all other occupied centres. After all, the workers were occupying the province's central communication network. In addition to British Columbia's intraprovincial telecommunications links, the occupiers handled telecommunications to Asia, Canada's west coast defence communication network, as well as national television and radio connections. Any attempt to isolate one exchange could have resulted in a complete communications shutdown for cities or even regions. To risk such a communications shutdown was unthinkable.

Before the extension of the occupation across the province, the company was in the process of seeking contempt charges against the union for the Nanaimo takeover. BC Telephone charged that since the December injunction specifically prohibited sit-ins, the mass occupation was in contempt of this injunction. In its defence the union argued that the occupation was provoked by the suspension of telephone workers in Duncan and Nanaimo. Morley Shortt, the union's lawyer, outlined the peaceful nature of the sit-ins and noted that there had been no damage to company equipment. The occupation had in fact defused the conflict. In an affidavit to the court Alf McGuire stated, "It is my opinion and the unanimous opinion of the executive of Local 3 of the Telecommunications Workers Union that we have averted a much more serious confrontation between the Union and the Company" and that the relations between the workers and lower management had improved as a result of the occupation.[27]

The court rejected the union 's argument and denounced the union for regarding itself as sole arbiter of "what is in the best interests of the public, the union members, and even of the company." Finding the union guilty of criminal contempt, the judge charged, "a more blatant affront to the authority of this Court, the law and of the basic principles of an ordered society would be difficult to imagine." The court ruled that the union would be fined an undetermined amount and that the fine would be increased for each day the union continued its occupation. Sentencing was suspended for two days, as the court awaited the union's response to the order that it evacuate the buildings.

The union had pledged to remain in the buildings until a contract was signed. Workers had joked about the union's policy of passive resistance. The following poem by a worker in the occupation was circulated throughout the BC Telephone buildings.

> T'was the night before an injunction
> And down at the "Boot"
> BC Tel was not sharing
> One cent of their loot.
> There I sat, telephone to my ear
> Awaiting the Mounties
> Who soon might appear,
> When all of a sudden someone opened the door,
> In passive resistance, I sank to the floor,
> There I lay, oh so short, so round and so chubby,
> Thinking of parents, children and hubby,
> Then I thought "what the hell" show no fears,
> Show no flusters,
> 'Cause to drag me out
> They'd need four men and a tow truck from Buster's.

After the court's decision the union leadership realized that they would eventually be forced out of the buildings. Discussion within the union leadership turned to whether or not the tactic of passive resistance should be followed. The executive felt that adhering to the tactic might divide the union, with some members remaining in the buildings to be carried out, and others walking out on their own. The solidarity, cooperation, and general good feeling built up during the occupation would be lost if some workers left the buildings out of fear of arrest, family commitments or physical intimidation. As well, the risk of confrontation with police might lead to damage and violence which would be blamed on the union. The union would lose the support it had enjoyed to date.

A second tactic—defiance of the court order—was discussed but the majority view was that this would lead to the smashing of the union. With the union convicted of criminal contempt, the TWU was no longer taking on just the telephone company. Defiance of the court order meant the union would have to contend with the police, the courts and the military—in short, the Canadian state.

Neither of these alternatives was considered realistic. The

Leigh Taylor

Pickets at 768 Seymour Street.

union, after all, was interested in signing a collective agreement. The occupation had challenged "the rights of property" but the goal was to force the company to settle the sixteen-month-long dispute, not to nationalize BC Telephone or establish workers' control. The executive decided that with the workers on the street and the public behind the union, the dispute could still be won. The executive ordered the occupiers out of the buildings.

On Monday evening, February 9, a few hours after the court decision, the union issued a press release and sent a communication to all occupied buildings. The statement commended the members for the occupation: "We have proven that our disciplined membership is willing and able to provide the people of British Columbia with telephone service despite countless management provocations designed to lock us out." It described the court ruling as "granting the company the lockout which BC Tel has not been able to achieve on its own," and promised escalation in the form of a province-wide strike. The statement included instructions to be followed during the evacuation of the buildings. Anticipating that the company would accuse workers of sabotage, the union instructed local areas to arrange for tours of all occupied buildings before they were vacated.

The union suggested that a police officer, a representative of BC Telephone management, a local labour leader, and the press should all be included on the tour. After establishing that no telephone equipment or facilities had been damaged, the workers were to leave the company premises *en masse* in a disciplined march out.[28]

Most of the buildings were vacated late Monday evening or early Tuesday morning. The one exception was 768 Seymour Street, BC Telephone's "nerve centre." The tour of the twelve-storey building started at 9 am and ended with a dramatic march out of the building at noon. Trade unionists, largely construction workers from downtown Vancouver, left their job sites shortly before noon and gathered in front of the building in a massive show of solidarity. The demonstration filled the street and crowded into a four-storey parking garage opposite the building. Hanging from the parking garage was a giant sign declaring, "Beastly Tel is alive and leeching." At noon the telephone workers marched out of the building led by a TWU member playing the bagpipes. It took a full twenty minutes for the workers to file out of the building. A brief rally was held on Seymour Street, with Jim Kinnaird promising the continued support of the BC Federation of Labour.

For the first few days of the strike, local areas sent flying pickets to shut down anything remotely connected with BC Telephone. The union leadership warned that the union was awaiting sentence in the contempt decision and that further violations of the injuction would leave the TWU in a precarious position. Local strike captains were asked to restrict picketing to two workers per building entrance. The return to the streets after the five-day occupation left emotions at a high level. During the occupation TWU members had worked long and hard to staff the exchanges. Once on the streets, they wanted to force an immediate settlement.

Speaking in parliament in Ottawa on the day that the occupation ended, NDP MP Mark Rose characterized BC Telephone's actions in the dispute as "union busting." Rose referred to BC Telephone as "some cross between a Southern textiles corporation and a nineteenth century Texas cattle baron" and asked labour minister Gerald Regan whether he was planning to intervene in the dispute. Regan replied that he regretted that BC Telephone had not accepted the Peck report and declared that the company "was perhaps not as sensitive to the situation in British Columbia as might be desired." That same day Regan appointed William Kelly, senior assistant deputy minister of federal mediation and conciliation services, as mediator in the dispute. Regan arranged for the union and company to meet with Kelly by the end of the week.

After six days of talks with Kelly, negotiations broke off again. The company demanded that any settlement be contingent upon a further telephone rate hike. The demand shocked mediator Kelly, who claimed that "we have an agreement, but I can't cope with a situation where one party [BC Tel] puts a third party [CRTC] into the picture." Kelly stated that the company's demand was "a new experience in any mediation I've been involved in." Booking out of the dispute, Kelly stated: "As far as I'm concerned, it's no longer a labour dispute, it's strictly a matter between the company and the regulatory commission." Labour minister Regan called the company's move "bizarre" and characterized the demand as "totally outside the field of labour relations . . . I am not aware in my experience of any occasion in history in which any utility company ever before thought to put such a clause in a collective agreement."[29]

The newspapers also condemned the company's proposal. The Vancouver *Sun* termed it "corporate blackmail" and charged that "with one crude slash the company has cut its own credibility in this dispute." The *Province* regarded the company's proposal as "preposterous," stating that "no company can expect a guaranteed recovery of its costs and such a suggestion can come only from someone dwelling in Never-Never Land." The Kelowna *Daily Courier* charged that "BC Tel has slapped the process of collective bargaining in the face" through its "ridiculous ploy."[30]

Bill Clark called the company's move "a stupid, dangerous and extremely provocative thing to do." Clark suggested that with the court suspension of sentence on the contempt convictions, the company was seeking to prolong the strike so as to further provoke workers. This latest move, explained Clark, "brought in a condition which neither of the two parties to the bargaining process were capable of fulfilling . . . BC Telephone revealed to everyone that it never intended to bargain in good faith." The union decided that along with its "aggressive picketing policy," it would lay charges against BC Telephone for failing to bargain in good faith.[31]

Toward the end of February negotiations were reopened with Michael Collins once again as mediator. On March 2 a tentative agreement was reached. A compromise was worked out on the jurisdiction and scheduling issues. On the two remaining money items, the company agreed to increase the accumulated time off from six days to twelve days per year effective 1982 and accepted a wage increase equivalent to Peck's recommendation.[32]

But the dispute was far from over. Over the previous month BC Telephone supervisors had fired 24 TWU members for strike-related activities. Most back-to-work agreements, including previous BC Telephone-TWU agreements, guarantee that no disciplinary action will be taken against employees as a result of the strike. The union regarded these firings as victimizations and insisted that the back-to-work agreement allow the full work force to return to work. Clark protested that the company's stance was a "travesty of justice," with BC Telephone setting "itself up as judge, jury and executioner." If the TWU allowed the company to get away with these firings, argued Clark, "every struck employer would simply fire strikers to weaken the union and break the strike."[33]

The company argued that the fired employees had "abused their strike privilege"—a statement which infuriated trade unionists, who felt that strike action was a right not a privilege. BC Telephone proposed that the union seek reinstatement of the workers through the grievance procedure. The company urged the rest of the strikers to return to work while the fate of the 24 was settled. This proposal was rejected by the union and on March 6 talks broke off once again.[34]

In the last week of February the BC Federation of Labour announced the details of its program of support for the TWU. One-day general strikes were to be held in different regions of the province. Federation president Jim Kinnaird warned that the one-day actions might culminate in a province-wide general strike. "We'll start exercising our rights with our feet," warned Kinnaird. "We'll start walking off the job from one end of this province to the other." Nanaimo—where the occupation had begun and a city noted for its strong labour traditions—was appropriately chosen as the centre for the first one-day general strike.[35]

On Friday, March 6, Nanaimo was shut tight for one day. Ferries, buses, Harmac Pulp and Paper Mill, Crown Zellerbach, MacMillan Bloedel, Assembly Wharves, SuperValu, Safeway, Overwaitea, construction sites, provincial government offices, liquor stores, federal government offices, post offices . . . every workplace where there was union was closed from midnight Thursday to midnight Friday. Members of the Pulp and Paper Workers Union of Canada, thought not affiliated to the BC Federation of Labour, joined the one-day protest action, with its local president, Bill Bryant, declaring, "we are with the Telecommunications Union 100 per cent."[36]

In place of the usual TWU picket signs, participants in the one-day general strike wore "BC Tel Crown Corporation Now" signs. Picketers handed out leaflets which explained the need to nationalize BC Telephone. The press condemned the solidarity action. Despite such criticisms the federation announced that a second solidarity strike would take place on March 20 in the East Kootenays, an important resource centre for mining and forestry.

A few days after the Nanaimo general strike, provincial NDP leader Dave Barrett put forward a recommendation to both the union and the company in an attempt to end the impasse. Barrett

André Bérubé

Trains, stores, ferries, buses, government offices — virtually everything — shut down during the Nanaimo sympathy strike. This sign says it all.

suggested that the union members return to work and that the union and company submit the fate of the 24 employees to a single arbitrator. But this suggestion was not acted upon because the company refused to allow the 24 employees to return to work pending the arbitrator's decision.

After the breakdown in discussions the provincial labour minister, Jack Heinrich, intervened behind the scenes with what was essentially Barrett's plan. Heinrich persuaded his former law partner, Allan Hope, to serve as sole arbitrator. On March 14 the union and the company agreed on a back-to-work arrangement. Subject to membership ratification of the contract, all employees were to return to work on March 23. The evening before the return to work, Hope was to submit a binding interim decision on the 24 fired workers. Hope had the power to suspend any or all of the 24 workers. Those suspended would report to work but leave immediately. They would collect their full salary pending the final outcome of the arbitration.[37]

During the following week the TWU held ratification meetings around the province. On March 20 the contract was adopted with 81 per cent in favor and the next day the agreement was signed. A few days earlier, with word of a possible settlement, the BC

Federation of Labour had postponed indefinitely its second one-day general strike.

On Monday, March 23, the telephone workers returned to work. Hope brought down his interim decision that ten employees were to be temporarily suspended. Ten days later in his final report, Hope ordered full reinstatement of the suspended employees. Hope explained that the strike had been remarkably free of violence. With 10,000 workers on strike, Hope commented that "the mathematics of the dispute indicate that there were hundreds of confrontations daily between union members and supervisors . . . I can say that there was not so much as a bloody nose in those hundreds of individual confrontations that took place." BC Telephone immediately announced that it would appeal the decision to the Supreme Court of British Columbia even though both sides had agreed that Hope's decision would be binding.[38]

After a confrontation lasting 536 days, including a four-month selective strike, a seven-day occupation in Nanaimo, a five-day province-wide occupation of telephone exchanges, a one-day general strike in Nanaimo, and the intervention of the federal labour minister, the provincial labour minister, the provincial leader of the opposition and half a dozen mediators, the telephone workers had concluded another collective agreement.

Postscript

The telephone workers were fined $60,000 on their contempt of court conviction. The union regarded the fine as excessive, in light of the peaceful nature of the action. The fine cost roughly $1 per member per day of the occupation. Most organized workers felt that the TWU had been fortunate, as the courts usually hand out jail terms for such confrontations and defiance—Canadian labour leaders have been jailed for a far lesser challenge to the rights of property.

BC Telephone won its appeal to the British Columbia Supreme Court on the Hope arbitration. Hope had contended that "during a legal strike, an employer has no right to dismiss or otherwise discipline striking employees," and that such actions would be "contrary to industrial relations principles." The recourse open to employers in the face of unlawful conduct by striking employees was "civil action for damages or an injunction or to lay criminal charges." The judge, however, ruled that Hope's contention amounted to "an error in law." As the case stands today, all 24 employees are back at work. The company seems content to have made its point of law while accepting the substance of the arbitration.[1]

In the wake of the 1980 dispute the company set improved employee relations as its priority goal. The company spent hundreds of thousands of dollars on television advertisements to assure its employees and the public that "BC Telephone is people." The advertisements featured various workers and explained how their jobs were necessary and socially valuable. The company initiated quality circles. These management-run discussion groups were designed to give workers the illusion of having power over decision-making in the workplace. Input from workers was encouraged, but group leaders were appointed by management to limit the scope of discussions. Experienced union members recognized the quality circles as a management tactic to undermine the union by encouraging workers to believe that problems in the workplace could be resolved through these informal discussion groups guided by management personnel.

In June, 1982, as the TWU met in convention, the company brought down an ultimatum: either the union would accept a rollback in pay and hours or the company would lay off workers. BC Telephone proposed that employees work a 6½-hour day—a more than 13 per cent cut. The supervisors had already been forced to make similar concessions through the loss of one work day every two weeks. As well, supervisors were pressured to come in and work without pay on the day they were scheduled to miss. BC Telephone called the program "job sharing." But as the union pointed out, the company refused to guarantee that there would be no layoffs if the union accepted the rollback. The union therefore rejected the scheme.[2]

After 1,800 permanent employees and 500 temporary employees received layoff notices, the union protested the action to the Canada Labour Relations Board on the grounds that the layoffs were a result of technological change, not economic conditions. The company assured the board that the layoffs were a temporary adjustment because of declining revenues caused by the current recession. When the union queried how the company could argue that the layoffs were indeed temporary, the company assured the board that it expected an upturn by the end of the year and that all employees would be recalled by February 4, 1983. The board ruled that the layoffs were legal and that the burden of proof was on the union to establish that the layoffs were not the result of economic causes. By implication, the company was suggesting that the "virtual economic disaster" which Gordon MacFarlane had said in 1977 would be necessary before any layoffs would be contemplated, had arrived.[3]

Following the CLRB hearing, the TWU took BC Telephone to court over the layoffs. Union lawyer Morley Shortt argued that the Canada Labour Code specifies that employees are entitled to sixteen weeks' notice when more than 300 employees are laid off. The company explained that regulations added to the code through a federal order in council permitted companies a shorter notice period than specified in the Canada Labour Code. The court sided with the union and ruled that the regulations which shortened the notice period were beyond the power of the cabinet and that the employees were to be reinstated and given notice of sixteen weeks as stipulated in the code plus two weeks as stipulated in the contract.[4]

The company reinstated the laid off employees in compliance with the court order. Their back pay, however, was placed in a trust fund pending the outcome of a company appeal to a higher court. BC Telephone then announced that it was laying off 2,100 employees as of December 2—1,365 plant employees and 735 clerical employees, but no permanent traffic staff. The company had stated before the CLRB in July that the cuts were temporary and that all employees would be called back no later than February 4. This time, BC Telephone changed its tune and referred to the cuts as "termination notice." Once again the company claimed the layoffs were based on "economic necessity." To the company, economic necessity meant that increases in profits had not reached projected levels; the term did not imply that it was not making profits.[5]

An obvious case can be made that the layoffs are timed to have an effect on bargaining. Negotiations on a new collective agreement opened in mid-September. With 2,100 employees facing the axe, the move could well be construed as an attempt to "soften" the union for the negotiations. But whatever the outcome of the negotiations, there is little reason for telephone workers to expect that things will improve at BC Telephone as long as it is a part of the GTE empire.

Since the sixties the TWU has called for the nationalization of BC Telephone. The constant turmoil of industrial relations at the company, the decline in service, and the massive profits of the foreign-owned monopoly utility make it a prime target for public ownership. Regulation has proven to be totally inadequate in forcing the company to meet the needs of the public or the workers.

Three issues are raised in opposition to public ownership of the telephone company; costs of purchasing the company would be prohibitive; nationalization of a federally-chartered company cannot be accomplished by the provincial government; and other public enterprises have failed. To determine the purchase price, or compensation, the government would have to undertake an intensive study of BC Telephone's financial affairs. This would necessarily include an examination of subsidiaries such as AEL Microtel, formerly Lenkurt Electric. The results might well reveal that the price of acquisition would be far below even the lowest estimates.

The fact that BC Telephone is outside of provincial jurisdiction could be overcome by joint action with the Quebec government. A GTE-owned holding company, Anglo-Canadian Telephone, which is incorporated in Quebec, owns BC Telephone as well as Quebec Telephone. By collaborating with Quebec, the holding company could be nationalized without the intervention of the federal government.

Finally, arguments which point to the inefficiency and problems of government-run services merely underline the need to include workers in the control and management of such enterprises. Nationalizations which result in government bureaucracies running the company offer no solution. Public ownership with control shared between elected representatives from the ranks of the telephone workers and directors elected by the general public would provided British Columbia with an attractive alternative.

The massive force which the union had to muster merely to conclude the 1980 contract does not bode well for employees of BC Telephone. The problem the TWU faces is much more than simply that of an arrogant employer, though BC Telephone has amply fitted that description. The crisis at the telephone company raises fundamental questions for our society. At issue is the assumption that management has the right to manage because what is good for big business is good for the nation.

Canada's adversarial system of industrial relations is described as a quest for balance between the powers of organized labour, the powers of management, and the public interest. But fundamentally all power rests with management. It has the power to hire, fire, discipline, expand, shut down plants, determine what will be produced, and how and where production will be organized.

The public can choose only to purchase or not to purchase the products or services of a particular firm. Often, however, there are no real choices. For example, while one can choose from a variety of automobiles, as long as there is no mass transit system the individual is forced to opt for some form of private transportation. With utilities the choices are even more limited. Here consumer choice amounts to have service or not to have it.

Workers have the choice of working under the conditions outlined by management or finding another job. In individual cases—through collective agreements, arbitration, or appeal to labour boards—workers can stall, reverse, or cushion the adverse effects of management decisions. But for the most part the role of labour codes and regulatory legislation is to prevent work stoppages during the life of collective agreements. Their function weakens unions and aids management.

The Vancouver *Province* charged that the 1981 TWU occupation of the telephone exchanges challenged the industrial relations system:

> The takeover destroys a fundamental assumption in labour-management relations in the western world. The understanding is that, while they may have different ideologies, each side has certain basic securities. Among these, the employer assumes that the property remains in his control, while the employee assumes he will be paid for his work.

The TWU, however, did not seek to overturn the fundamental assumptions of the labour-management relationship. Nevertheless, there is no question that its actions constituted a challenge to the powers of management. As far as telephone workers were concerned, they were doing their usual work and providing the services they had always provided. What they were challenging was the presumption that management has the right to force workers out onto the streets, either by lockout or layoffs. Increasingly the TWU and other unions faced with technological change will be forced to challenge the belief that management's absolute control of the production of goods and services is legitimate.

The assumption is that management gains its rights through its investment of capital. Yet the owners are generously rewarded for their financial investment. Workers too invest, but their contribution—the actual production of goods and services—is far more valuable than the compensation they receive in the form of wages, benefits, and pensions, and ill deserves the insecurity of layoffs, lockouts, or the presumptions of an arrogant employer.

This challenge to the so-called fundamental assumption of the labour-management relationship is part and parcel of the history of the labour movement and particularly the history of the telephone workers. The battles for recognition saw management refuse to meet with workers' representatives because they viewed unions as interfering with their right to manage. Today, with the constant threat of technological change, the survival of unions and labour organizations depends on workers asserting their rights to control in the workplace.

Notes

Chapter 1 *Early Years*

1. John Brooks, *Telephone: The First Hundred Years,* Harper and Row, New York, 1975. p. 71.
2. William Patten, *Pioneering the Telephone in Canada,* privately printed, Montreal, 1926, p. 123.
3. Robert Collins, *A Voice from Afar,* McGraw-Hill Ryerson, Toronto, 1977, p. 128.
4. E. B. Ogle, *Long Distance Please: The Story of the TransCanada Telephone System,* Collins, Don Mills, 1979, p. 150. *British Columbia's Telephone Tapestry,* published by BC Telephone, (no date). Collins, op. cit., p. 129.
5. *Daily British Colonist,* Victoria, January 22, 1880.
6. G. W. S. Brooks, "Edgar Crow Baker: An Entrepreneur in Early British Columbia," MA thesis, UBC, 1976.
7. Ibid.
8. BC Telephone Historical Record, Chapter 1880 (b), pp. 33-34. City of Vancouver Archives. These billing features were symptoms of the fact that the industry was in its infancy. It had not yet been decided whether telephone service would ultimately be billed by a toll on the amount of time used, as long distant rates are calculated today, or by subscription for limitless use, the way local calls are handled.
9. Lindsay Allen, "Factors in the Development of the British Columbia Telephone Industry, 1877-1930," MA thesis (early draft), SFU, 1981, pp. 14-19.
10. *Telephone Talk,* "Early history of the BC Telephone Company is revealed," Vol. 14, No. 11, November 1924.
11. Allen, op. cit., pp. 19-20.
12. *Telephone Talk,* op. cit., p. 12. Included in the amended act were a number of clauses forbidding the use of Chinese labour. Up until this time the company had occasionally used Chinese workers on construction. The inclusion in the act

of a penalty of up to $25 for every Chinese employed effectively excluded these workers from telephone work for a number of years. The legislature added the discriminatory measure at the urging of the construction foreman, who insisted, reflecting the racism of the day, that Chinese workers were more careless than white workers.

13. Allen, op. cit., pp. 30-40.
14. Ibid., pp. 49-64.
15. *Telephone Talk,* "E. J. Haughton, of Victoria, was a boy telephone operator," Vol. 27, No. 2, March/April, 1937. BC Telephone Historical Record, Chapter 1883 (c), pp. 1-4.
16. BC Telephone Historical Record, Chapter 1888 (c), pp. 1-6.
17. *Telephone Talk,* Vol. 12, No. 3, March 1922.
18. *Telephone Talk,* "A talk with the man who erected the eighty-foot poles," Vol. 16, No. 5., May 1926.
19. Ibid.
20. There are many articles in *Telephone Talk* on telephone work in the early years, but three are particularly valuable: "When the Kootenay started and the telephone was in demand," Vol. 8, No. 8, August 1919; "When the telephone was a curiosity in the capital city," Vol. 12, No. 2, February 1922; and "Telephone policy progressive from the beginning," Vol. 12, No. 5, May 1922.
21. Michael A. Mulcaire, *The International Brotherhood of Electrical Workers,* publisher unknown, Washington, DC, 1923.
22. Ibid, pp. 1-4. *History and Structure of the IBEW,* IBEW (1980).
23. Mulcaire, op. cit., pp. 4-11.
24. *Livewire,* "The First 80 Years," Vol. XXXIII, No. 1, November 1981.
25. Mulcaire, op. cit., pp. 32-37.
26. *Telephone Talk,* Vol. 8, No. 2, February 1918.
27. BC Tel. Historical Record, Chapter 1902 (g), pp. 1-3.
28. Ibid., pp. 3-21.
29. *Daily Province,* "Telephone bell still tinkles," November 26, 1902.
30. *Daily Province,* "Telephone strike is still unsettled," November 26, 1902.
31. Ibid.
32. *Labour Gazette,* January 1903, pp. 566-567.
33. *Daily Province,* "First he would and then he wouldn't," December 1, 1902.
34. *Daily Province,* "News-Advertiser discredited at board of trade," December 3, 1902.
35. *Daily Province,* "Renew efforts to end strike," December 5, 1902.
36. *Labour Gazette,* January 1903, p. 517.

Chapter 2 *The Employers' Offensive*

1. Paul Phillips, *No Power Greater: A Century of Labour in BC,* BC Federation of Labour and Boag Foundation, Vancouver, 1967, pp. 41-48.
2. Michael Bliss, *A Living Profit: Studies in the Social History of Canadian Business, 1883-1911,* McClelland and Stewart, Toronto, 1974, pp. 74-94.
3. *Labour Gazette,* May 1905, p. 12188.
4. IBEW 230 Minutes, August 14, 1905.
5. IBEW 213 Minutes, September 28, and October 26, 1905. BC Telephone Historical Record, Chapter 1905 (k), pp. 1-2.
6. *Daily Province,* "No overtures for peace in strike," February 23, 1906.
7. BC Telephone Historical Record, Chapter 1906 (e), pp. 8-15. *Labour Gazette,* March 1906, pp. 1030-1031; July 1906, p. 100.
8. *Daily Province,* "Telephone girls may go on strike this afternoon," February 21, 1906; "Telephone girls are on strike," February 23, 1906.

9. BC Telephone Historical Record, Chapter 1906 (e), pp. 1-24.
10. *Daily Province,* op. cit., February 23, 1906; "Telephone strike is still in effect," February 24, 1906.
11. *Daily Province,* op. cit., February 23, 1906.
12. BC Telephone Historical Record, Chapter 1906 (e), p. 22.
13. *Daily Province,* "Labor's support to the girls," March 27, 1906, p. 1.
14. *Daily Province,* "Line workman is heavily fined", June 8, 1906.
15. *Daily Province,* "Striker's camera was smashed by 'phone official'," March 17, 1906; "Assault cases dismissed," March 20, 1906; "Superintendent Kent fined," March 20, 1906.
16. *Labour Gazette,* July 1906, p. 100.
17. One of the problems in establishing how the strike ended, and when exactly BC Telephone was removed from the unfair employer list, is that the minutes of Local 213 from October 1906 to December 1908 are missing. The next book of minutes, starting in January 1909, clearly show that by this time the operators' auxiliary had ceased and that IBEW linemen were again employed by BC Telephone.
18. IBEW 213 Minutes, January 1909 — Feb. 1912.
19. *Labour Gazette,* December 1911, pp. 567-571.
20. Department of Labour, Trades Disputes Record, PAC RG 27, file 2335. *BC Federationist,* March 21, 1913, p. 1.
21. Frederick Winslow Taylor, *The Principles of Scientific Management,* W. W. Norton, New York, 1967 (1911). For an excellent study on "scientific management" see Harry Braverman, *Labour and Monopoly Capital: The Degradation of Work in the Twentieth Century,* Monthly Review Press, New York, 1974.
22. *Telephone Talk,* Vol. 2, No. 5, May 1912.
23. *BC Federationist,* July 31, 1914, p. 5. *Labour Gazette,* April 1913. pp. 1079-1080.
24. *Telephone Talk,* Vol. 4, No. 8, August 1914.
25. Ibid.
26. Ibid.
27. *Telephone Talk,* Vol. 3, No. 1, January 1913.
28. *Telephone Talk,* Vol. 3, No. 5, May 1913.
29. *Telephone Talk,* Vol. 2, No. 11, November 1912.
30. *Telephone Talk,* Vol. 3, No. 10, October 1913.
31. BC Telephone Historical Record, Chapter 1914 (k), p. 1.
32. *Labour Gazette,* October 1916, pp. 1663-1666.
33. *Labour Gazette,* November 1917, pp. 899-907.
34. BC Telephone Historical Record, Chapter 1918 (c), p. 1.
35. Interview with Gordon Farrell by BC Telephone historian A. P. (Tony) Farr, October 18, 1976.
36. Copy of the illuminated address, BC Telephone Library, Burnaby.
37. It is not clear why the operators' local was called 77A. Local 77 was the Seattle local of the IBEW. The operator organizer who helped form the Vancouver operators' local was from Seattle, but it is not known whether the operators' local in that city was also called 77A, and in fact whether the two operator locals were linked. For most of its brief history 77A acted more as a sub-local of Vancouver Local 213 than as an independent local directed by the autonomous operators' department.
38. Michael A. Mulcaire, *The International Brotherhood of Electrical Workers,* Washington, DC, 1923, pp. 32-37.
39. *BC Federationist,* June 7, 1918, p. 8.
40. *BC Federationist,* July 5, 1918; July 12, 1918.

Chapter 3 *General Strike*

1. Michael A. Mulcaire, *The International Brotherhood of Electrical Workers,* Washington, DC, 1923, p. 28.
2. Mulcaire, op. cit., pp. 16-20. IBEW Local 213 Minutes, January 1909—June 1914.
3. IBEW 213 Minutes, July 27 and August 3, 1914. Records the total vote in favor of a general strike — 28 with 6 opposed. The total membership of the local was approximately 400.
4. *Live Wire,* IBEW 213, November 9, 1951, p. 22.
5. Stuart Marshall Jamieson, *Times of Trouble: Labour Unrest and Industrial Conflict in Canada 1900-1966,* Task Force on Labour Relations, Study No. 22, Ottawa, 1968, pp. 167-168. *Live Wire,* op. cit. p. 23.
6. *BC Federationist,* "Labour temple scene of trouble and rioting," August 9, 1918. Not much is known about Foxcroft. She was the operator of the PBX at the Labour Temple and was listed as the last secretary of the operators' Local 77A in 1921 (see Annual Report, BC Dept. of Labour, 1921).
7. H. A. Logan, *Trade Unions in Canada: Their Development and Functioning,* MacMillan of Canada, Toronto, 1948, pp. 306-313.
8. David Jay Bercuson, *Confrontation at Winnipeg: Labour, Industrial Relations and the General Strike,* McGill-Queen's University Press, Montreal, 1974, pp. 132-141. *Labour Gazette,* June 1919, Vol. 19, No. 6.
9. *BC Federationist,* June 6, 1919. *Live Wire,* op. cit., p. 24.
10. *BC Federationist,* May 30, 1919.
11. *BC Federationist,* June 6, 1919.
12. *BC Federationist,* June 13, 1919.
13. Department of Labour, Canada, Trade Disputes Record, strike file 19(285), Public Archives of Canada, RG27 2338. BC Telephone Historical Record, Chapter 1919 (g), pp. 1-3.
14. *Vancouver Citizen,* June 16, 1919, No. 1. Some copies of this newspaper of the Vancouver Citizens' League are available at UBC Special Collections. BC Telephone Historical Record, Chapter 1919 (g), pp. 1-3. Interview with Leona Copeland, November 10, 1981.
15. *BC Federationist,* "Hello girls are not easily beaten," July 11, 1919. *Live Wire,* op. cit., p. 25.
16. Leona Copeland, op. cit.
17. *BC Federationist,* "General strike called off at 5 pm Thursday," July 4, 1919.
18. Department of Labour, op. cit. The men's return to work a day before the women has become a legend in the TWU. I have interviewed operators from 18 to 80 years of age, and they have consistently asked me about "the strike where the men sold out the women." What is remarkable is that even today operators hear about this strike. It is part of the union's folklore handed down by generations of operators.
19. Paul Phillips, *No Power Greater: A Century of Labour in BC,* BC Federation of Labour and Boag Foundation, Vancouver, 1967, pp. 82-84.
20. *BC Federationist,* "Electricians win injunction," March 26, 1920. *Live Wire,* op. cit., p. 25. The internationals saw the 1919 strike in Winnipeg and the sympathetic strikes as initiated by the OBU.
21. *BC Federationist,* "Another union loses charter," August 29, 1919. *Live Wire,* op. cit., p. 25.
22. *BC Federationist,* "Electricians win injunction," March 26, 1920.
23. Department of Labour, op. cit., pp. 1-2.
24. IBEW 213 Minutes, July 11—December 15, 1921.
25. *BC Federationist,* "Rest room for 'hello' girls," September 26, 1919. *Live Wire,* op. cit., p. 25.

26. *Telephone Talk,* "The system now has over ninety thousand telephones," Vol. 16, No. 6, June 1926, pp. 18-19.
28. *Telephone Talk,* "British Columbia leads Canada in telephone saturation," Vol. 16, No. 8, August 1926, p. 20.
28. E. B. Ogle, *Long Distance Please: The Story of the TransCanada Telephone System,* Collins, Toronto, 1979, pp. 44-46 and 56.
29. *Telephone Talk,* "How the telephone bill was paid twenty years ago," Vol. 16, No. 7, July 1926, pp. 10-13.
30. *Telephone Talk,* "PBX serves a variety of interesting purposes," Vol. 14, No. 9, September 1924, pp. 5-10.
31. *Labour Gazette,* June 1927, pp. 612-614.
32. Ibid., p. 614.

Chapter 4 *Company Union*

1. Lindsay Ross Allen, "Factors in the Development of the BC Telephone Industry 1877-1930," MA thesis (early draft), SFU, 1981. An "independent" was any telephone company not owned by the massive Bell network.
2. Gary set up a number of holding companies and was continually reorganizing his interests. While the telephone operating company maintained the name BC Telephone Company throughout, Gary set up British Columbia Telephone and Telegraph Corporation for the purchase of the operating company and then switched the holding company's name to National Telephone and Telegraph Corporation a few months later. National was reorganized as Gary's Canadian holding company, Anglo-Canadian Telephone Co. Ltd., in May, 1935. Today this is BC Telephone's immediate parent, a holding company for GTE.
2. Ibid., pp. 71-73.
3. Ibid., p. 70.
4. E. B. Ogle, *Long Distance Please: The Story of the TransCanada Telephone System,* Collins, Toronto, 1979, pp. 123-136.
5. Robert Collins, *A Voice from Afar: The History of Telecommunications in Canada,* McGraw-Hill Ryerson, Toronto, 1977, p. 208.
6. Allen, op. cit., pp. 162-167.
7. Ibid., p. 75.
8. Ibid., pp. 75-115.
9. Patricia Roy, "Regulating the British Columbia Electric Railway: The First Public Utilities Commission in British Columbia," *BC Studies,* No. 11, Fall 1971.
10. Allen, op. cit., pp. 151-153.
11. Ibid., pp. 168-175.
12. Ibid., p. 182.
13. Interview with Art Marsden, March 26, 1981.
14. Interview with Roy Thompson, March 25, 1981.
15. Memorandum of Agreement, BC Telephone Co., North-West Telephone Co., Kootenay Telephone Co. Ltd., Chilliwack Telephones, Ltd., and Mission Telephone Co. Ltd., and BC Electric Employees' Organization, dated August 1, 1930.
16. Interview with Frank Owston, March 24, 1981. Robert Collins, op. cit., p. 218.
17. E. B. Ogle, op. cit., pp. 62, 77.
18. Victoria Branch Minutes, February 2, 1931. Vancouver Branch Minutes, December 21, 1932. Victoria wrote to the company over a worker who had been laid off in spite of four years' seniority over anyone else in his department. The company refused to explain the layoff, but assured the branch that the worker would be the first rehired employee.
19. Victoria Branch Minutes, March 2, 1931 and June 6, 1932. Interview with Jake

Barker, March 27, 1981. Interview with Frank Owston, March 24, 1981.
20. Victoria Branch Minutes, October 26, 1932.
21. Agreement of Working Conditions and Schedule of Wages, BC Telephone Electric Employees Organization, November 12, 1934.
22. Victoria Branch Minutes, June 5, 1933, November 6, 1933.
23. BC Telephone Company Electrical Employees' Organization, General and Local By-Laws of the Victoria Branch, 1935.
24. *Telephone Talk,* Vol. 28, No. 2, March/April 1938, pp. 5-7.
25. *The Telephone Worker of British Columbia,* Vol. 1, No. 8, April 1937.
26. Vancouver Branch Minutes, February 7, 1938; February 18, 1938.
27. *Telephone Talk,* "Our operating school reopened in Nineteen Thirty-Five," Vol. 26, No. 1, p. 12, January 1936.
28. Directorate Minutes, September 1940. Vancouver Branch Minutes, October 30, 1940.
29. Ogle, op. cit., p. 186.
30. *Telephone Talk,* "Behind the lines," Vol. 33, No. 2, March/April 1943.
31. Directorate Minutes, September 12, 1942.
32. *Telephone Talk,* Vol. 34, No. 4, July/August 1944.
33. Paul Phillips, *No Power Greater: A Century of Labour in B.C.,* BC Federation of Labour and the Boag Foundation, Vancouver, 1967, pp. 128-129.
34. Quote by Milner, Vancouver Branch Minutes, April 5, 1943; June 5, 1943.
35. Vancouver Branch Minutes, June 5, 1943.
36. Directorate Minutes, April 10, 1943. Vancouver Branch Minutes, June 7, 1943. Minutes of Council of Telephone Operators' Organization of BC, Vancouver, May 18, 1943. The operators' local was first incorporated under the Societies Act of British Columbia, but later, like the EEO, sought recognition under PC 1003.
37. *Telephone Talk,* "Behind the lines," Vol. 33, No. 2, March/April 1943.
38. Work Agreement, Telephone Operators' Organization of British Columbia, October 1, 1943.
39. Ibid., Article 13, clause (d).
40. Minutes Joint Meeting Council and Directorate, May 8, 1944.
41. Minutes FTW Executive Council, February 15, 1945.

Chapter 5 *A National Union*

1. Jack Barbash, *Unions and Telephones: The Story of the Communications Workers of America,* Harper and Brothers, New York, 1952. "Toward Industrial Unionism: Bell Telephone Workers and Company Unions, 1919-1937," *Labour History,* No. 16, Winter 1975.
2. Conference Report, *Meeting of the Telephone workers held in Pender Auditorium,* Vancouver, British Columbia, November 4-8, 1946.
3. Material on the Interior BC Telephone Operators' Union can be found in the CLC collection, Public Archives of Canada, Ottawa, MG 28 I 103, Vol. 96, CLC.
4. National War Labour Board, Reason for Decision, Re: Okanagan Telephone Company (Vernon, BC) and Interior BC Telephone Operators' Union; also Letter to Regional War Labour Board from Interior BC Telephone Operators' Union, May 25, 1944 and Letter to A. R. Mosher, President, CCL, from Dan O'Brien, June 30, 1944, PAC, MG 28 I 103, Vol. 96, CLC.
5. Ibid., p. 4.
6. Constitution of the Independent Brotherhood of Telephone Workers, Winnipeg, Manitoba, January 1, 1947.
7. Interview with Ralph Wyatt, August 11, 1980.
8. Conference Report, op. cit.
9. Ibid., Tuesday morning Session.

10. Ibid., Thursday, Friday Sessions.
11. Proposed Constitution for the "Canadian Federation of Telephone Workers."
12. Correspondence re Amalgamation 1946-1948. Traffic Council Meeting Minutes, May 4, 1948. Minutes of Meeting with Okanagan Telephone Employees for the Purpose of Joining with the FTW of BC (Plant Division) September 10, 1949.
13. *Federation New and Views,* Number 25.
14. "News and Views," Number 25, September 1949. "A Little History," No. 2., "The Sub-Committee's campaign to put the Federation into the CWA," No. 3.
15. Board of Directors' Sub-Committee Meeting with Management, July 21, 1949.
16. Letter from C. B. Diplock to D. J. Weaver, October 4, 1949.
17. Ballot Results, December 1, 1949; Summary of votes on Affiliation.
18. Semi-Annual Meeting of the Board of Directors, July 9, 1951.
19. Working Agreement and Schedule of Wages, Federation of Telephone Workers of BC and BC Telephone Company, June 1, 1948. Starting with the 1952 agreement, there were two separate wage scales: one designated by numbers 1 through 9 and the other designated by letters A through E. The vast majority of male clerical workers were classified under the letter wage scale and women under the number system. The letter scale had a significantly higher base rate and ceiling. For example, a group 5 daily rate varied from a starting wage of $7.31 to $9.96 after five years. A group C daily rate varied from $8.95 to $14.27 after five years.
20. For example, the company boasted during the conversion to dial in downtown Vancouver in 1941, that "Telephone operators on the permanent staffs of Seymour and Trinity did not lose their jobs with the elimination of those offices." *Telephone Talk,* "All of downtown Vancouver now served by dial," Vol. 31, No. 3, May/June 1941.
21. Working Agreements, FTW and BC Telephone: 1946, 1947, 1948, 1950 and 1952.
22. *Telephone Talk,* "3 Times more 'phones being installed than before war", Vol. 38, No. 5, September/October 1948.
23. The earliest versions of the automatic dial switching equipment installed in British Columbia in the forties and fifties had in fact been patented in 1891, by Almon B. Strowger, a Kansas City undertaker. Strowger invented the automatic switching equipment after he became convinced that his business was losing customers because operators were not properly forwarding calls and were telling his customers that he was out or busy. The solution for Strowger was to eliminate the need for operators to handle calls. The patent for Strowger's switch was the foundation upon which the Automatic Electric manufacturing company was built.
24. John Brooks, *Telephone: The First Hundred Years,* Harper and Row, New York, 1975, p. 193.
25. *Daily Province,* "BC Telephone will employ married women," October 5, 1946, p. 5. *Telephone Talk,* "Automation + telephony = more people," Vol. 48, No. 1, January/February 1958.
26. *Telephone Talk,* "Distance dialing for BC in 1959," Vol. 47, No. 3, May/June 1957. "Traffic, staff metering-instantly," Vol. 49, No. 6, November/December 1959.
27. *Telephone Talk,* "Microwave path testing completed in BC," Vol. 45, No. 5, September/October 1955. "Groundwork laid for microwave in BC," Vol. 46, No. 3, May/June 1956.
28. *Telephone Talk,* "Directory delivery now 'mechanized'," Vol. 46, No. 3, May/June 1956; "No coffee breaks for this secretary!" Vol. 45, No. 5, September/October 1955.
29. *Telephone Talk,* "All Bayview telephones 'cut' to Cedar-Cherry," Vol. 38, No. 5, September/October 1948. "Automation + telephony = more people," op. cit.

Chapter 6 *A Time of Change*

1. The actual figures for job losses between the fifties and the early sixties is difficult to establish, particularly from membership records. Layoffs, early retirement, resignations, transfers, and attrition accounted for a reduction of between 1,500 and 3,000 workers. *The Plant Division Bulletin,* June 1961, estimates that 1,500 wage earners were eliminated from the payroll. A federal Department of Labour conciliation officer estimated the membership as 8,000 in 1959 and 4,000 in 1962 indicating the loss of 4,000 jobs. BC Telephone annual reports quote 7,130 employees in 1957 and "under 5,330" in 1962, placing the job loss at 1,800. 1959 Conciliation Records, PAC, RG 27, Vol. 1826, file 760.604.59 and 1962 Conciliation Records, PAC, RG 27, Vol. 1834, file 760.838.62.
2. Working Agreement between FTW and BC Telephone Co., April 1, 1960, p. 41.
3. *Plant Division Bulletin,* June 1961, Vancouver.
4. Plant Council Minutes, April 17-19, 1961.
5. Interview with Don Bremner, May 31, 1982. Interview with Larry Armstrong, May 31, 1982.
6. Interview with Bert Johns, December 16, 1980.
7. *Short Circuit,* Vol. 1, Nos. 1-7, September 1962 through March 1963. *"The 'Short' Short Circuit",* May 1963 edition.
8. Plant Council Minutes, April 6, 1963.
9. *Plant Division Bulletin,* May 1963. Plant Council Minutes, June 10-15, 1963. Special Meeting, Board of Directors, September 5, 1963. The eight members were Harvey Cooper, Jack Sagris, Cyril Francis, Richard Murray, Maurice Anderson, Thomas Stanley, George Spence, and Richard Langlois.
10. Interview with Robert Baillie, June 2, 1982.
11. Interview with Bert Johns, op. cit.
12. Winnipeg Conference, November 1-3, 1957. Near verbatim notes by Jim Armour, Canadian Telephone Employees Association; and notes by author unknown (one of the CWA delegates), Conference of Telephone Workers, Toronto, February 12-14, 1958. Letter from CWA District 10 director John Lax to Norm Pettigrew, November 29, 1960. *Clerical Division Bulletin,* Fall 1962.
13. According to the Traffic Council Minutes January 20, 1959, the approximate number of operators in 1957 was 3,000, and in 1958 2,700. Minutes for June 6, 1960 show traffic workforce at BC Telephone 1,567, OK Telephone 63, total 1,630. Minutes for June 22, 1962 show BC Telephone 1,097, OK Telephone 36, total 1,133.
14. Interview with Elizabeth Lillquist, June 1, 1982.
15. Minutes, Special Meeting of the Traffic Council, February 11, 1958. The council comprised the general secretary and five councillors from Vancouver Local 10 and one councillor each from Victoria Local 11, Nanaimo Local 12, Nelson Local 13, Cafeteria Workers Local 14, Kamloops Local 15, and Okanagan Local 16.
16. Traffic Council Minutes, June 16, 1959; June 6, 1960; January 23-24, 1961; and June 12, 13, 1961. Interview with Lillquist, op. cit.
17. Traffic Council Minutes, September 20-21, 1962. Board of Directors Minutes, October 3, 1962. *CWA News,* "Canadians on CWA school: good show," December 1962, Washington DC. It is not clear why clerical advised against the seven hour day. Aside from the money, one other possibility is that the operators switched from a one hour meal break to half hour meal break (in both cases unpaid).
18. *CWA News,* op. cit. Vancouver *Sun,* " 'Company union' cry stirs storm," May 26, 1965. *Clerical Division Bulletin,* Fall 1962.
19. Traffic Council Minutes, January 21-22, 1963; June 10-11, 1963. Letter from Martin L. Levinson, barrister and solicitor, to William Dunn, April 26, 1963. *CWA Cross Talk,* District 10, Toronto, July 26, 1963. *CWA BC Telephone Organizing*

Committee Newsletter, May 3, 1963; August 5, 1963.
20. Traffic Council Minutes, September 13, 1963. *Plant Division Bulletin,* October 1963. *CWA Cross Talk,* September 20, 1963.
21. Federation Telephone Workers of British Columbia, Fall 1963, signed by Norm Gravel.
22. Writ between Lillquist and B. H. Johns, J. N. Gravel and Hazel Magee, November 7, 1963. *CWA Newsletter,* November 1963.
23. CLRB, CWA, applicant, and BC Telephone, respondent (Traffic Division), File 766:1464:63, November 25, 1963.
24. Traffic Council Minutes, January 20-22, 1964; June 15-17, 1964.
25. Lillquist attributes the failure to the tactical error in starting in the outlying areas and to company interference in favor of the FTW. Interview with Bert Johns, op. cit.
26. *BC Telephone Organizing Committee Newsletter,* November 20, 1963.
27. Vancouver *Sun,* op. cit, May 26, 1965.
28. Ibid. Board of Directors Minutes, June 1964. Traffic Council Minutes, June 15-17, 1964.
29. GTE, "Background Information from GTE," Fact Sheet No. V, reissued by Corporate Communications Department, March 1979.
30. *The Province,* "BC Telephone control acquired by General," June 22, 1956. Interview with Bert Johns, op. cit.
31. Interview with Lillquist, op. cit.
32. Letter from D. S. Tysoe to Bernard Wilson, June 8, 1959. RG 27 Vol. 1826, File 760.604.59.
33. BC Telephone, "Code of Ethics," signed J. E. Richardson.
34. Interview with Art Marsden, March 26, 1981.
35. Vancouver *Sun,* "BC Telephone makes top profit," February 24, 1966.
36. Stuart Marshall Jamieson, *Times of Trouble: Labour Unrest and Industrial Conflict in Canada, 1900-66.* Task Force on Labour Relations, Study No. 22, Ottawa, 1968, pp. 458-459.
37. Ibid., pp. 419-422.
38. Paul Phillips, *No Power Greater: A Century of Labour in BC,* BC Federation of Labour and Boag Foundation, Vancouver, 1967, p. 155. *FTW Bulletin,* "Re: Oil Strike," November 29, 1965.
39. The BC Telephone employees fell under the Canada Labour Code, but the employees at OK Telephone and CT and S were under the British Columbia Code.

Chapter 7 *First Strike in Fifty Years*

1. Letter from D. S. Tysoe to W. P. Kelly, director, Conciliation and Arbitration Branch, Dept. of Labour, January 21, 1969.
2. Tysoe took over from J. R. Currie as conciliation officer in the 1962 dispute and had subsequently been assigned in the 1964, 1966 and 1968 disputes between the FTW and BC Telephone. Vancouver *Sun,* "Phone staffs halt work to rule," January 11, 1969.
3. *The Province,* "Phone wage proposal rapped," May 14, 1969.
4. Interview with Bert Johns, December 16, 1980.
5. Letter from D. S. Tysoe to W. P. Kelly, May 26, 1969.
6. Vancouver *Sun,* "Charges hurled by phone union," June 5, 1969. "Telephone official denies firm using 'scare tactics'," June 6, 1969.
7. Vancouver *Sun,* "BC Tel posts guards, clamps down on security," June 6, 1969.
8. Ibid. "The union's answer to BC Telephone company's statement," June 7, 1969.
9. Vancouver *Sun,* "A report to telephone customers," June 6, 1969.

10. Vancouver *Sun,* "The union's answer to BC Telephone company's statement," June 7, 1969.
11. Vancouver *Sun,* "Rotating strike plan approved," June 13, 1969.
12. *The Province,* "BC Tel warns employees to avoid rotation strikes," June 10, 1969. Letter from Tysoe to W. P. Kelly, May 26, 1969.
13. *The Province,* "BC Tel strikes today," June 23, 1969. "Strikers pull plug on all of BC Tel," June 27, 1969. Vancouver *Sun,* "Phone strike hits all BC," June 27, 1969.
14. *The Province,* "Tel plugs gaps," June 28, 1969. The strikebreakers included 1,300 supervisors and management personnel and approximately 200 employees (most of them Rand formula non-union employees and students hired for summer work) who crossed picket lines.
15. Vancouver *Sun,* "Some service disruptions as phone workers strike," June 23, 1969.
16. Vancouver *Sun,* "Students at BC Tel condemned," July 7, 1969.
17. Vancouver *Sun,* "1.3 million lost in wages, BC Tel says of strikers," July 5, 1969; "Building trades back phone union," July 10, 1969. *The Province,* "Phone strike extended to 'hot' building sites," July 10, 1969.
18. Vancouver *Sun,* "Court backs pickets in phone strike," July 18, 1969; "Phone union permitted to picket job," July 22, 1969.
19. *The Province,* "Join strike, supervisors urged," July 11, 1969.
20. R. B. Worley, *The Wonderful World of W. A. C. Bennett,* McClelland and Stewart, Toronto, 1971, p. 236. Victoria *Daily Times,* "Why stop with BCE?," February 27, 1962. *The Province,* "Berger laughs, hits charges," July 26, 1969.
21. *The Province,* "Mediator revived tel talks," July 24, 1969.
22. *The Province,* "Tel talks run past midnight," July 31, 1969. Vancouver *Sun,* "BC Tel, union reach accord," July 31, 1969.
23. Interview with Phil Zander, November 24, 1981. Interview with Theresa North, June 29, 1982. Interview with Bert Johns, December 16, 1980. Interview with Doug Booth, June 29, 1982.
24. Vancouver *Sun,* "Phone workers return to jobs," August 5, 1969. "Court orders pickets off SFU project," July 19, 1969.
25. Interview with Bob Donnelly, June 24, 1982. Interview with Dolly Storey, June 29, 1982.
26. Interview with Jim Bremner, November 18, 1981.
27. Vancouver *Sun,* Jack Wasserman column, September 11, 1969.
28. *The Province,* "Minor squabbles hit BC Tel," August 6, 1969.
29. Ibid.
30. Of the two fired employees who faced legal charges arising out of the incident, one was convicted of assault and the other acquitted, see Vancouver *Sun,* "Man jailed for assault," September 10, 1969. The acquitted employee, David Patterson, was eventually reinstated. Interview with Bert Johns, December 16, 1980. Vancouver *Sun,* "BC Tel employees refuse overtime in fight to back man fired in '69," September 2, 1972; "Phone worker dispute ends," October 19, 1972.
31. Vancouver *Sun,* "Strike points up job dullness," August 13, 1969.
32. Vancouver *Sun,* Jack Wasserman column, September 11; October 10, 1969.
33. Letter from Boris Mather to Bert Johns, August 25, 1969. The separation of the Canadian locals from the CWA was initiated by the international. Shortly after the failure of the raid on the FTW, the CWA convention passed a motion to withdraw from Canada. The Canadian locals were encouraged to set up the Canadian council as a transitional body which would act as an intermediate step to a full break with the international. The founding convention of the new Canadian union — the Communications Workers of Canada (CWC) — was held in April, 1972.
34. *The Federation Paper,* CFCW, December 1973.
35. Plant Council Minutes, June 1-5, 1970. *Federation Paper,* CFCW, February 1974.

36. Plant Council Minutes, General Secretary's Report, June 1-5, 1969. Vancouver *Sun*, BC Tel advertisement, August 22, 1969.
37. Working Agreement between FTW and BC Telephone Company, effective January 1, 1971.
38. Letter from D. H. Cameron, Chief Conciliation Officer to J. S. Gunn, Acting Director, Conciliation and Arbitration Branch, Department of Labour, Ottawa, January 5, 1973. Vancouver *Sun*, "BC Tel hints at rate boost," February 10, 1973. *The Transmitter*, Clerical Division, "Editorial: The Ethics of Bargaining," January 1973.
39. Vancouver *Sun*, "Phone operators working to rule," January 9, 1973. "BC Tel executive quits," January 30, 1973. *The Province*, "BC Tel post filled," January 31, 1973. Bruce was made a vice president in 1970.
40. Vancouver *Sun*, "Conciliators named," February 16, 1973. Conciliation Board Report Dispute between BC Telephone Company and the FTW of BC, signed by Hugh G. Ladner, Chairman, Donald E. Taggart, Union Nominee, May 4, 1973, also Report of Company Nominee, signed by George S. Cumming, May 4, 1973.
41. *The Province*, "BC Tel contract signed," June 12, 1973.
42. Interview with Doug Booth, June 30, 1982. Collective Agreement between FTW of BC and BC Telephone Company (with plant and traffic appendices), signed June 11, 1973. Collective Agreement between FTW of BC and BC Telephone Company (with clerical appendices), effective January 1, 1973.
43. *The Transmitter*, August 31, 1973.
44. *The Transmitter*, November, 1973.
45. Vancouver *Sun*, "450 walk off Jobs at Okanagan Tel," August 7, 1973.
46. Vancouver *Sun*, "4,000 out at BC Tel," August 23, 1973. *The Transmitter*, August 1973.
47. Special Plant Council Meeting, August 29, 1973.
48. *The Transmitter*, November 1973. Vancouver *Sun*, advertisement, FTW, October 13, 1973.
49. *The Transmitter*, November 16, 1973.
50. *The Transmitter*, November 29, 1973.
51. *The Transmitter*, January 1974. Interview with Bob Donnelly, June 30, 1982.

Chapter 8 *Technological Change*

1. Tom Forrester, ed., *The Microelectronics Revolution: A Complete Guide to the New Technology and Its Impact on Society*, MIT Press, Cambridge, Massachusetts, 1981. *Microelectronics*, a Scientific American Book, W. H. Freeman and Company, San Francisco, 1977.
2. John Johnson, testimony, CRTC Hearings 1980.
3. Linda Rolufs, testimony, CRTC Hearings 1980.
4. Rod Hiebert, testimony, CRTC Hearings 1980.
5. *The Transmitter*, June 24, 1976.
6. Plant Council Minutes, January 15-19, 1973, General Secretary's Report.
7. *The Transmitter*, March 1973, July 12, 1974.
8. *The Transmitter*, February 13, 1974. *FTW Bulletin*, June 3, 1974.
9. "Don't be short-changed on service by BC Tel," FTW, 1974. BC Telephone, *Management Bulletin*, Vol. 17, No. 11, February 20, 1974.
10. Vancouver *Sun*, "Phone men walk out in protest," February 22, 1974. Letter from J. B. Lidster, director of industrial relations, BC Telephone to R. G. Donnelly, president, FTW, August 1, 1975. *The Federation Paper*, "Arbitrator rules for company on Phone Mart," November 1974.
11. Vancouver *Sun*, "Phone workers walk out," April 23, 1974. Nanaimo *Daily Free*

Press, "No service cut in BC Tel strike," April 23, 1974. Working Agreement between BC Telephone and FTW, January 1, 1973, pp. 9, 54, and 60.
12. Vancouver *Sun,* "College students in phone dispute branded 'scabs' by labour council," April 29, 1974. Nanaimo *Daily Free Press,* "Scab labour charge denied by BC Tel," April 29, 1974.
13. Special Plant Council Meeting, May 3, 1974.
14. Vancouver *Sun,* "Walkout spreads at BC Tel," June 25, 1974.
15. *The Province,* "10,000 Workers to pull BC Tel plug," June 27, 1974.
16. Vancouver *Sun,* "9,000 BC Tel Workers launch wildcat strike," June 27, 1974.
17. *The Transmitter,* "Service centre training: the whole story," July 12, 1974. Vancouver *Sun,* "BC Tel loses in dispute," July 10, 1974.
18. Vancouver *Sun,* "BC Tel service 'has deteriorated,' panel told," October 29, 1974.
19. Vancouver *Sun,* "BC Tel 'favours regulation' by provincial government," May 29, 1975. *Financial Post,* "This is BC Tel's answering service," April 6, 1974. Vancouver *Sun,* "Barrett bid to elect pair to BC Tel board defeated," March 13, 1975.
20. Vancouver *Sun,* "BC Tel man grilled on equipment purchasing," October 22, 1974. "BC Tel supplier affiliated company," October 23, 1974. *The Province,* "CTC urged to 'free' BC Tel purchases," July 11, 1975.
21. Vancouver *Sun,* op. cit., October 29, 1974.
22. Vancouver *Sun,* "Decision pledged 'soon as possible' on BC Tel rate revision application," November 2, 1974.
23. Vancouver *Sun,* op. cit., October 22, 1974. November 2, 1974.
24. Vancouver *Sun,* "BC Tel relationships to be probed by Ottawa," February 15, 1975. The February 1975 rate increase was 4.6 percent on basic residential and 9.5 percent on commercial subscribers.
25. *The Province,* "BC Tel 20% bid 'shocking,' " March 15, 1975.
26. Ibid.
27. Report of Conciliation Commission, in Dispute between BC Telephone Company and the FTW of BC, signed Hugh G. Ladner, February 21, 1975.
28. *The Province,* "Clerical workers at BC Tel claim they are working to rule," March 11, 1975.
29. Collective Agreement between BC Telephone Company and the FTW of BC, January 1, 1975.
30. Vancouver *Sun,* BC Tel trying to provoke walkout, unionist charges," May 30, 1975. "BC Tel signs agreement after two-month wait," June 19, 1975.
31. *Management Bulletin,* BC Telephone, Vol. 18, No. 14, February 19, 1975. Vancouver *Sun,* "170 laid off at BC Tel," April 18, 1975. "Bell accused of intimidation," July 10, 1975. *The Province,* "BC Tel to lay out record $192 million," February 13, 1974.
32. Vancouver *Sun,* advertisement, August 24, 1974.
33. Interview with George Yawrenko, July 14, 1982.
34. *The Province,* "BC Tel plans appeal," July 5, 1975. Vancouver *Sun,* "BC Tel loses appeal in layoff case," July 24, 1975.
35. Canada Labour Relations Board, Reasons for Decision, TWU and BC Tel and TEMPO, August 22, 1979, p. 66.
36. Ibid., pp. 20-3.
37. Ibid., pp. 30-31, 65-66.
38. Letter from J. Lorne MacDougall chief executive officer, CLRB, to Doug Boother, general secretary, clerical division, June 29, 1971. MacDougall declared, "It is quite right, as you stated in your application of May 12, 1971, that the wording of the Board's certificate issued on February 23, 1949, is sufficiently broad to include additional classifications initiated by the company after the date of the said Order of certification providing both parties are in mutual agreement regarding such inclusion. Where the parties fail to agree that the

incumbents of newly created classifications are employees within the meaning of the Act and hence within the meaning of the order of certification, the bargaining agent has two options to open to it, namely (a) the filing of a new application for certification under Section 7 or Section 8 of the Act, and (b) the filing of a request for review pursuant to the provisions of subsection (2) of Section 61 of the Act.

. . . Under either option, the Board's policy is to ascertain whether the applicant trade union has as members in good standing a majority of the employees comprising the group which it wishes added to the unit and in respect of whom it seeks the right to bind them by collective agreement."

39. CLRB, op. cit., p. 32.
40. Report of Conciliation Board in Dispute between FTW and BC Telephone, May 4, 1973, pp. 2-3. Ibid., pp. 34-35.
41. CLRB, op. cit., November 9, 1979.
43. CLRB, Certificate of Result of Vote, Re: Ballot by Mail of Certain Supervisory, Professional and Managerial Employees of the BC Telephone Company, file 535-1204.

Chapter 9 *The Long Distance Feeling*

1. Federation Council Minutes, February 8, 1977. CLRB, Order of Certification, May 5, 1977.
2. Plant Council Minutes, January 21-26, 1974.
3. C. R. Schenk, "Massey Workers and Wage Controls: A Case Study of Mobilization and Coercive Integration," Draft PhD thesis, University of Toronto, 1982.
4. Vancouver *Sun,* "250 BC Tel workers walk out," February 24, 1976. Report of the Board of Arbitration, BC Telephone Company and FTW, April 15, 1976.
5. *The Province,* "BC Tel spells out job security from A to Z," advertisement, November 26, 1977. MacFarlane replaced Basil Beneteau in February 1977 after Beneteau resigned and moved to Northern Telecom Ltd.
6. Vancouver *Sun,* "BC Tel ad misleading, union charges," November 28, 1977.
7. Letter from Gordon MacFarlane to fellow employees, September 23, 1977.
8. Conciliation Commissioner's Report in a Dispute between the BC Telephone Company and the Telecommunications Workers Union, Dr. Noel A. Hall, May 31, 1977.
9. Arbitration Report, BC Telephone Company and FTW, October 27, 1976.
10. Vancouver *Sun,* "BC Tel dispute report treated warily," June 7, 1977.
11. Negotiation and Disputes Notes, compiled by Don Frost.
12. Vancouver *Sun,* "Protest trucks snarl traffic in BC Tel clash," September 21, 1977. 768 Seymour Street had been BC Telephone's headquarters until the building at 3777 Kingsway (known disparagingly as the "Boot") was built.
13. Interview with Theresa North, July 28, 1982. *The Province,* "BC Tel strikers hunt scab workers," October 13, 1977. The account in *The Province* is a brief report from company spokesperson Keith Matthews. My sources are from interviews with participants. North regrets her own and the operators' inexperience at the time of the walk-through. With hindsight, she suggests that they should have sat-in once they entered the building.
14. *The Province,* "Phone crews on job today?" October 14, 1977.
15. Vancouver *Sun,* "Phone services to be cut but bills will still arrive," November 25, 1977.
16. *The Province,* "BC Tel spells out job security from A to Z. 9,165 times," advertisement, November 26, 1977. "BC Tel jobs ad labelled false," November 28, 1977.

17. *The Province,* op. cit., November 28, 1977. Vancouver *Sun,* op. cit., November 28, 1977.
18. *The Province,* "Something is out of order at BC Tel," November 24, 1977.
19. *Catch-22,* published in 1961, is a novel by Joseph Heller. The title is his apt description of a dilemma in which the victim cannot win.
20. Vancouver *Sun,* "Police investigating phone tie-up 'plot,' " December 3, 1977.
21. Vancouver *Sun,* "Early slowdown forecast following 'hot' edict on BC Tel," December 24, 1977.
22. *The Province,* "Judge enters BC Tel dispute," January 11, 1978.
23. Vancouver *Sun,* BC Tel pickets step up mass harassment," January 12, 1978.
24. Vancouver *Sun,* "Judge probes chances of BC Tel agreement," January 16, 1978.
25. Vancouver *Sun,* op. cit., January 12, 1978.
26. Interim Report of the Industrial Inquiry Commissioner in the Dispute between BC Telephone Company and the TWU, Mr. Justice H. E. Hutcheon, January 24, 1978. In the interim draft Hutcheon accidentally deleted the specification that all contracting out should go to "an available recognized union firm whose work is satisfactory." This phrase was subsequently included.
27. Vancouver *Sun,* "BC Tel settlement reached," February 3, 1978.
28. *The Province,* "Tel union can return but . . .," February 10, 1978. Vancouver *Sun,* "BC Tel getting ready for normal work Monday," February 11, 1978.
29. Vancouver *Sun,* "BC Tel employees back on the job as contract signed," February 13, 1978. *The Province,* "Long, ruthless battle seen at BC Tel," September 24, 1980.
30. *The Province,* "Tel workers vote 91.4 percent for 3 year pact," February 13, 1978.

Chapter 10 *The Occupation*

1. Vancouver *Sun,* "Sorry, wrong number," March 26, 1980.
2. Tom Stanley, originally from Toronto, joined the company in 1957 as an apprentice installer in Vancouver. He was an active member of Local 1 and was one of the eight Local 1 dissidents tried in the early sixties. In the late sixties he moved to Prince George, returning to the coast in 1970 after being elected assistant general secretary in plant. The union sent him back to Prince George for a year when it decided to establish a union office there. In the mid-seventies Stanley was loaned to the CWC to assist in organizing Bell plant workers.
2. TWU Convention Minutes, Report of the One Union Concept Committee, June 11-16, 1979. The goal of the one union movement is far from complete. The vast majority of TWU locals are still composed almost entirely of either traffic, clerical, or plant workers. While the divisions no longer exist in name, in practice there are still wide differences between workers in the three work classifications. The move towards one union is viewed as a continuing process.
4. TWU Convention Minutes, June 11-16, 1979, Tactical Committee Report.
5. *BC Telephone Bulletin,* Vol. 23, No. 45, March 24, 1980.
6. *BC Telephone Bulletin,* Vol. 23, No. 77, May 23, 1980. TWU Press Release, May 23, 1980.
7. Conciliation Commissioner's Report, Dispute between BC Telephone Company and the Telecommunications Workers Union, August 27, 1980, pp. 1-9.
8. Bill Clark defeated incumbent Bob Donnelly at the June, 1980 convention. As president, Clark replaced Donnelly as union chief negotiator. TWU Press Release, September 11, 1980.
9. *BC Telephone Bulletin,* Vol. 23, Nos. 121, 122.
10. Vancouver *Sun,* "Rebels end phone centre sit-in," October 20, 1980.
11. Appendix to TWU Intervention to the CRTC Hearings on BC Telephone Company's Request for a Rate Increase, July 31, 1980.

12. Bob D'Etcheverrey, testimony, CRTC Hearings 1980.
13. Jim Gordon, testimony, CRTC Hearings 1980.
14. Russ McNicol, testimony, CRTC Hearings 1980.
15. Appendix to TWU Intervention, op. cit.
16. Vancouver *Sun*, "Don't blame workers for BC Tel rules," May 13, 1980.
17. *BC Telephone Bulletin*, Vol. 23, No. 192, December 3, 1980.
18. TWU Press Release, December 12, 1980.
19. Vancouver *Sun*, "Union claims Tel lock out," January 22, 1981; "85 workers sent home," January 27, 1981.
20. Telecom Decision CRTC, 81-3, British Columbia Telephone Company general increase in rate, 29 January 1981, p. 15-16.
21. Vancouver *Sun*, "No excuse for poor service now, says union chief," January 31, 1981.
22. Vancouver *Sun*, " 'Desperate' workers seize BC Tel office," February 4, 1981. Nanaimo *Times*, "BC Tel 'under new management'," February 5, 1981.
23. Vancouver *Sun*, "BC Tel sit-ins sweep province," February 5, 1981.
24. TWU Press Release, February 5, 1981.
25. Courtesy of Ron Jeeves, Local 3, Nanaimo.
26. BC Federation of Labour, Press Release, February 7, 1981. *The Province*, "Recipe for anarchy," February 10, 1981.
27. BC Telephone Company and Telecommunications Workers Union, Reason for Judgement, C804526, Vancouver, February 9, 1981.
28. TWU Press Release, February 9, 1981.
29. Vancouver *Sun*, "Regan rips 'bizarre' BC Tel," February 19, 1981.
30. *The Province*, "Preposterous proposal," February 19, 1981. Vancouver *Sun*, "Corporate blackmail," February 20, 1981. Kelowna *Daily Courier*, "BC Tel move is incredible," February 20, 1981.
31. *The Province*, "BC Tel demand shocks mediator," February 19, 1981. Vancouver *Sun*, "Union to seek court aid over BC Tel labour practices," February 21, 1981.
32. The compromise on the jurisdiction provided for a committee of six people, three from the union and three from management, to prepare job descriptions for plant. Disagreements would be sent to a sole arbitrator. A letter of understanding regarding scheduling was signed outlining a procedure where disagreements on scheduling would be resolved by agreement between the company's industrial relations manager and the TWU vice president or their appointees. If no agreement was reached, the issue would be resolved by a sole arbitrator.
33. Bill Clark, letter to the membership, March 5, 1981.
34. *The Province*, BC Telephone advertisement, March 8, 1981.
35. Calgary *Herald*, "BC labour group planning rotating general walkouts," March 4, 1981. *BC Fed Labour News*, Vol. 2, No. 1, March 1981.
36. Vancouver *Sun*, "Regional walkout brings Island area to standstill," March 6, 1981.
37. Arbitration Award, Allan Hope, BC Telephone and TWU, March 24, 25, 26, 27, 30, 1981. If the arbitrator decided to suspend or dismiss the employee, the union would have to reimburse the company for the employee's wages.
38. Hope's award only applied to 23 members. One of the fired employees, Mort Johnsen, had brought a libel suit against BC Telephone after he was fired on charges of destroying company property. The company later apologized, explaining that it was a case of mistaken identity. On receipt of a written apology Johnsen dropped the suit.

Postscript

1. Supreme Court of British Columbia, Reason for Judgement, No. A811723, August 20, 1981.
2. Both plant and clerical employees work a 7½-hour day; the proposed cut was to reduce the work day to 6½ hours. For traffic, with its odd scheduling, operators work 7 hours on day shift and less on evening and night shifts. The company never clarified how this scheme would apply to traffic.
3. *The Province,* "BC Telephone spells out job security from A to Z," advertisement, November 26, 1977.
4. Supreme Court of British Columbia, Reason for Judgement, No. A822240, August 5, 1982.
5. *BC Telephone Bulletin,* August 18, 1982.

Index